Mathematics for Computer Students

Second edition

Rex Wilton

 BLACKWELL

Copyright © Rex Wilton, 1992, 1996

The right of Rex Wilton to be identified as author of this work has been asserted in accordance with the Copyright, Design and Patents Act 1988.

First published 1992
Second edition published 1996

2 4 6 8 10 9 7 5 3 1

NCC Blackwell Ltd
108 Cowley Road
Oxford OX4 1JF
UK

Blackwell Publishers Inc.
238 Main Street
Cambridge, Massachusetts 02142, USA

Library of Congress Cataloging in Publication Data
A CIP catalogue record for this book is available from the Library of Congress

ISBN 1 85554 6671

British Library Cataloguing in Publication Data
A CIP catalogue record for this book is available from the British Library

Typeset in 10 on 12pt Palatino and Helvetica
Printed in Singapore

Mathe
Computer Students

Contents

Preface to the second edition   vii

Introduction   ix

Chapter 1   An introduction to modelling   1

Chapter 2   Constructing models   26

Chapter 3   Sets and their applications   45

Chapter 4   Number sets and systems   65

Chapter 5   A first look at number systems   76

Chapter 6   Mistakes and errors   94

Chapter 7   Storing and manipulating numbers within the machine   113

Chapter 8   Graphs, functions and inequalities   124

Chapter 9   Matrices and graphics   150

Chapter 10 Data structures   170

Chapter 11 Numerical and algebraic methods   204

Chapter 12 Diminishing uncertainty – statistics   230

Chapter 13 Will it work? Probabilities   257

Chapter 14 Applied two-state logic and switching   274

Answers to the exercises   301

Index   335

For Diana, my wife, who has given constant support and encouragement, and for my son, Petroc, who has read and experimented with much of the manuscript in the ultimate consumer test.

Preface to the second edition

This book has been many years in the making. Years during which I have taught similar subject matter on computing courses at all levels from school-leaving examinations to post-graduate and from initial computer-industry training to in-service professional courses. It has become increasingly apparent that few books exist which deal with what is truly relevant to the needs of computer students. Certainly vast areas of traditional mathematics courses are *not*.

It has been equally apparent that few people writing mathematics textbooks for computer studies courses realize that for the computer practitioner the demands are for real, robust programs which work and for real systems which are efficiently planned, costed, implemented and maintained. These systems must be able to deliver the very wide range of information types and quantities which the users of today need. The newcomer to the industry may find him- or herself contributing to the programming of computer-controlled machine tools, aircraft flight simulators or 'traditional' commercial data-processing. Equally he or she might be working as a member of a team of operators within a large mainframe system or as the sole systems/programmer/operator 'expert' in a small-business system. There are, in all these, common mathematical concepts and processes. A virtual common 'core' of essential mathematical understanding and skills.

From the moment of starting to prepare for joining the industry the new recruit has quite specific needs which can be met from the great toolbox of mathematics. All the individual elements are to be found within any modern mathematics syllabus in the schools but are rarely linked together, much less to the needs of computing. Yet much of so-called 'modern' mathematics is, indeed, in the syllabus precisely *because* of the development of computers. What is often lacking is any meaningful connection between the teaching in school and the *practical needs* of the real world of computing.

In this book I have attempted to bring together the computer practitioner's needs and the mathematics concerned. To provide the connective tissue, as it were. I have ruthlessly excluded any mathematics which does not have a clear and direct relevance.

There is much which is relevant for students on any computer course although the book has been written primarily for post-school, entry-level computer courses. These include the NCC's Threshold Course and its International Diploma, the BTEC Diploma and similar initial studies. In particular I have tried to write a book which may do something to help the many people who wish to work in computing but who have had little prior success with mathematics.

For those of us who need to use mathematical tools in practical ways the subject should not be shrouded in mysticism nor should we be 'process-bound'. Equally I believe that ordinary users should not be subjected to 'symbol-shock' whenever they open a book on the subject.

Users have been making ever-louder demands for 'user-friendly' computer systems. We, as important users, have just as much of a right to a user-friendly mathematical toolbox.

There is no such thing as a totally original book on any subject and I must acknowledge my indebtedness to pupils and students, spread over 30 years of courses, who have helped, by their experience of difficulties, to concentrate my thinking about both the mathematics and the teaching methods.

Numerous colleagues, in professional computing and in teaching, have also stimulated my thinking. George Penney and Rob Roseveare at NCC, Dr John Roberts and a former and much-respected teaching colleague, Alan Shaw, have all contributed in different degree to the development of the ideas embodied in this book. Since the publication and use in schools and colleges of the first edition of the book I have also benefited from talking with many teachers around the world and with their students. The feedback from this source has been invaluable and very positive. Many have warmly welcomed this more practical view of mathematics in computing. Needless to say the people I have mentioned are emphatically *not* responsible in any way for any imperfections or omissions in the text!

This book is not in any way intended to be a general mathematics course. It is rather a topic-by-topic exploration of a specific range of mathematical ideas and techniques and how and where we find them applied in our industry.

There are exercises within the text as well as at the end of each chapter, but if the topics are truly to be mastered plenty of practice within real applications is the only way which I know of developing true confidence.

Rex Wilton

Introduction

Mathematics means many different things to different people. Many people use highly specialized maths in their chosen trades or professions but few are *professional* mathematicians. We all use some elementary mathematics in everyday life yet quite excellent computer people may come from non-mathematical disciplines. What then are the *mathematical* needs of the computer student?

It is the author's belief, based on many years of working with both computers and mathematics, that a genuine understanding of a range of mathematical ideas and some competence in selecting appropriate mathematical processes together provide a key element of good computing practice. This applies to the operations staff, seeking to establish the reliability and the true capacity of their systems; to the programmer who needs to write efficient, effective and easily maintained software, or to make best use of existing packages, without having, each time, to 're-invent the wheel' and to the analyst, seeking to devise logical, efficient and cost-effective systems.

Many people, often as a result of early failure and poor teaching, have a deep distaste for mathematics, and may even fear it. If you are one of those people do not be frightened off just yet. Almost all of us have, in our natural intelligence, a capacity for mathematical thinking and practical competence which may, sadly, just have been 'blocked off' by unhappy early experience. In this case it takes a great deal of courage to start to look again at the subject. The hope of the author is that even the so-called non-mathematicians *will* take such a look, in the interest of becoming better computing practitioners.

Computing *was*, in earlier times, dominated by mathematicians and scientists merely because they, for centuries, had sought to devise large and fast calculating machines to help them in what most of them disliked in any case, the drudgery of elaborate calculations. If truth be known few people, even among mathematicians and scientists, have ever actually enjoyed 'doing sums'. It is a very tedious, low-grade activity to which the machine is much better suited than are people since *it* is very fast and rarely makes 'mistakes', nor does it ever get bored!

Since the transistor evolved into the amazingly small 'chip' the machines are now very much easier to use and so reliable that anyone can play! Moreover we can 'play' with physically small machines which have vast memories, high processing speeds and powerful software. In exploiting the power of the machine, but also in understanding its limitations, imposed by its own fundamental characteristics, some mathematical 'know-how' is invaluable.

It was fashionable at one stage to talk of computers as 'electronic' brains, but it is safer to think of them as 'high-speed morons'. They will do the wrong thing as readily as the right, introduce error where there was none and do the stupid as readily as the seemingly intelligent. They are ignorant of right or wrong, good or bad, accurate or inaccurate. *We* must supply these things and monitor what the machines *do* at all times.

Essentially the machine consists of myriads of switches, any one of which can only be 'on' or 'off' and this establishes what the machine can do. If 'on' represents a '1' or 'true' and 'off' represents a '0' or 'false' the machine can have a set of such switches arranged so as to **add numbers written as 1 and 0** and to **compare statements as 'true' or 'false'**. This is **all** the machine *can* do in its very core in the Arithmetic and Logic Unit (ALU). (It should be noted that this is not the noun, 'arithmetic', meaning sums, but the adjective, *arithmĕtic*, meaning *by the process of addition*).

Fortunately 'two-state' number and logical systems already exist among the interesting, but very specialized, playgrounds of professional mathematicians. These met practically all the new demands created by programmable computers and little further mathematical invention was needed in order to make the two-state transistor into an effective basis for the modern digital computer.

For us ordinary mortals the difficulty is that *everything* which we want the computer to process for us has to be converted, under very strict mathematical rules and laws, into this two-state arithmetic or 'strict' logic. The converted data must then be processed, also under largely mathematical rules, to produce the information which we need.

Fortunately the conversion to binary arithmetic is, these days, done for us by the interpreters or compilers but *we*, however, must establish the strict logic of our systems and software and ensure that the processing is done according to the rules. There is no help for it, we *must have* some grasp of mathematics!

Another problem is that the machine will introduce error in the process of conversion from everyday numbers to their binary equivalents. For example, 2.6 is an exact amount in our 'ordinary' numbers but cannot be translated into an exact amount in binary. Furthermore most of our numbers are far less 'exact' than we might think and in the vast volumes of data processed and the very high processing speeds in the computer such inaccuracies can be generated and propagated very quickly and very widely unless we control the processing.

The final and most important thing which we must understand is that the computer cannot process *reality*. The real world must be converted into numerical and logical descriptions which *can* be processed or stored in binary form.

Mathematics is a truly international language for *describing* many aspects of the real world. Its symbols and conventions are virtually standard throughout the modern world. At the heart of every computer program is a representation of some aspect of reality in this language. We are accustomed to such representations of reality in many other contexts; we call them 'models', small imitations of the real thing. Mathematical models are, most commonly, *abstract* structures formed of symbols and there are long-established processes both for constructing and for manipulating them. It is this modelling which lies at the heart of computing.

This book is devoted to taking a fresh look at mathematics as it relates directly to good computing practice. We shall consider, step-by-step, the skills of creating the mathematical models, look at the existing processes for manipulating those models, numerical and logical, and at the special demands imposed by the machine's two-state systems. We shall also be concerned, throughout, with the control of the processing and the safeguarding of acceptable levels of accuracy.

1

An introduction to modelling

Objectives

By the time you have worked through this chapter you should be able to:

- ❑ understand the basic principles of modelling
- ❑ construct numeric, symbolic and spatial models
- ❑ understand the operation of the laws of mathematics
- ❑ see the place in computing of logical and statistical models
- ❑ construct simple models in pseudo-code.

1.1 Numerical models

1.1.1 The simplest of all mathematical models are built up with ordinary numbers which describe the **number** or the **size** of something, according to need. A herd of cows in a field may seem simply a pleasant country scene but to the farmer or a scientist that scene may be a complicated picture in which there will be many different things, depending upon the level of detail at which he or she needs to investigate.

If we merely observe such a scene we may never even need to put it into words. If, however, we need to **communicate** or **store** the details of what we are seeing we must **describe** what we see in some way. Just how we do this will depend upon *what* we need to convey and *to whom*. To an artist or an author this could be a very complex task but he or she will still find it impossible to represent everything in a scene and will **select what is essential** to the story or the picture in order to create the necessary 'atmosphere'.

1.1.2 The process of mathematical modelling is similar but even more selective. It is a process of continuing to **eliminate** *all irrelevant* **information**, no matter how interesting it may be, until **we can simplify our description no further and no possible ambiguity remains**. For a mathematical model we might simplify the country scene, described above, by making it absolutely clear that we are concerned *only* with the cows and with nothing about the cows except how many there are. 'Ten cows' is then the simplest possible model.

1.1.3 This number itself, as written in figures, is a model developed at a much earlier stage in history when the numerals 1, 2, 3, 4, 5, 6, 7, 8, 9 were invented to record the number of fingers upon which the cows were counted, and the cypher, 0, to indicate when and where there is nothing (or no cows). Supplement these numerals, or figures, with a column system and we have a model-making set which can describe, in the simplest possible way, any quantity no matter how large or how small. All that we need in addition to the numbers is, in everyday life at least, a label to show 'of what?'. The number, 10, for instance is a purely theoretical quantity until we attach to it, in our example, the label 'cows'. The label is vitally important, not only in showing what we are **'enumerating'** (giving a number to or counting), but because, as we shall see later when we consider error, the 10 when applied to cows is significantly different from the same number in *10 metres*.

1.2 Symbolic models

1.2.1 If we advance one step further and contribute some symbols to denote the operations which we can *perform* upon numbers, +, −, ×, ÷ (commonly called 'operators') and a symbol for equivalence, '=', we are able to make much more useful models which are only slightly more complicated. (In computing practice we use '*' instead of '×' and '/' in place of '÷' and this convention will be followed in this book). We must also be aware that the '+' and '−' symbols each have two meanings. They may be used as an *instruction*, or operator, as in 4 + 2, or 4 − 2. Their second use is in *describing* a number value as in +2 or −2, that is to say two more than zero and two less than zero respectively. Some authors have tried using the signs in a different position to indicate the adjective, $^{+}2$ or $^{-}2$ but while this does emphasise the difference it has not become general practice.

1.2.2 The manipulations which we may perform with numbers and operators are not, as any people think, new tricks which have to be learned for each different application. They are always consistent and are governed by a very small number of laws. Since these laws must be observed in all our modelling we should **study** them and learn them thoroughly.

We will add a few others to the list as we meet the need for them but the great majority of our processes are governed by the first ten, which are listed below.

1.3 The laws

1.3.1 **(1a)** $a + b = b + a$ **(1b)** $a * b = b * a$

(2a) $a + (b + c) = (a + b) + c$ **(2b)** $a * (b * c) = (a * b) * c$

(3) $a * (b + c) = (a * b) + (a * c)$

(4a) $a + (-a) = 0$ **(4b)** $a * (1/a) = 1$

(5a) $a + 0 = a$ **(5b)** $a * 1 = a$

(6) $a * 0 = 0$

1.3.2 In all these statements we are using a, b and c to stand for any quantities we may chose to replace them by. You should practise replacing the letters with number values in each of the laws in turn and satisfy yourself that each law always works. (You may have noticed that the operations of subtraction and division do not figure in the list; why?)

It may well be the first time that you have seen the laws set out in this way, but you will have been obeying them for years if you have been getting the right answers to your calculations. If you have tended to get a lot of answers wrong it may be because you had no idea that such laws existed and have been doing things a bit haphazardly.

Some may look very obvious, but in any case we should have a little more knowledge than just the bare list so let us give the names of the laws and their interpretation.

1.3.3 (1a) and (1b). The **commutative** laws for addition and multiplication.

The word 'commutative' means 'in both directions, as in the word 'commuter', one who travels *to* work, in one direction, and *back from* work in the opposite direction. Mathematically it means that **if we are *only adding or multiplying* a set of numbers**, (no matter how many, even though only two are shown above), **the sequence in which we add or multiply those numbers does not affect the result.** One traditional way of checking our working, if we have 'added up' a column of numbers, is to add them 'down' and an equally effective way of checking $7 * 3$ is to do also $3 * 7$.

In practice it also means that we may put such a sequence of additions or multiplications into a model in the order in which they are most easily understood even if this means re-arranging a traditional way of writing the model. Check for yourself what happens with subtraction and division. Are they also 'commutative'?

1.3.4 (2a) and (2b) The **associative** laws for addition and multiplication.

This is a variation on the sequencing of operations which, among other things, allows us to split a sequence in different ways if it is easier to do so or if it helps confirm the accuracy of what has been calculated. For example 15 + (17 + 5) may be easier for us as (15 + 5) + 17. Once again you should practise picking out the easier associations of numbers. Try testing the law to see if it is equally true for subtraction and division.

1.3.5 (3). The **distributive** law for multiplication.

This is a little more complicated but it tells us that if we are **to multiply a number which is made up of a two or more other numbers added together we can multiply each of the parts and add them all together at the end.** It is this law which governs our 'long' multiplication. We could speak of 7 * 432 as 'seven times four-hundred-and-thirty-and-two' or 7 * (400 + 30 + 2) and we do, in fact use the commutative law to change it to 7 * (2 + 30 + 400) and perform it as (7 * 2) + (7 * 30) + (7 * 400). We could still set out multiplications in this way but we have, over the centuries, also learned some 'carry' procedures which make the whole thing look rather more complicated.

A combination of laws allows us to perform 297 * 463 as

$$(200 + 90 + 7) * (400 + 60 + 3) \text{ or}$$
$$7 * (3 + 60 + 400) +$$
$$90 * (3 + 60 + 400) +$$
$$200 * (3 + 60 + 400).$$

Explore this and similar examples and try to identify which laws govern which stages of the whole operation.

1.3.6 (4a) and (4b) The laws of **inverses** for addition and multiplication.

These, although they may seem a little strange at first, actually do crop up in many applications and may be explained in two slightly different ways. For addition:

(i)   **For every number ('plus' number) there is an exact opposite (negative or 'minus') number which gives the answer 0 if the two are added together,** *or*, if sum of two numbers is 0 the numbers must be inverses of each other.

(ii)   If we add to any number its own (additive) inverse the result is zero.

$$2 + {-2} = 0$$

For multiplication:

(i)   **for every number there is another number which generates the answer 1 if the two numbers are multiplied,** *or*, if, when two numbers are multiplied the answer is 1 the numbers must be inverses of one another.

(ii)   If we multiply inverses the result, (product), is 1.

Following on from these two laws we have a very important development. We can achieve the same result, for any addition or multiplication, by performing the 'opposite', (inverse) operation with the inverse value.

For instance: $2 \div \frac{1}{3} = 2 * 3$. Similarly $4 - (-2) = 4 + 2$. This is a process which we use very frequently in simplifying models.

1.3.7 (5a) and (5b). The laws of **identity** for addition and multiplication.

These two laws may look so obvious as to need no discussion, but we do use them extensively in modelling and the idea of something 'retaining its identity', i.e. remaining unchanged, is one upon which we often rely. Briefly we can sum up these laws as meaning that **if the effect of any operation we may perform, no matter how complicated, is to add nothing to a number, or to multiply that number by 1, the number will remain unchanged.**

In simplifying fractions by 'cancelling' for example we might, from 28/44, derive the equivalent form of

$$\frac{7 * 4}{11 * 4}$$

which we can then re-write as

$$\frac{7}{11} * \frac{4}{4}$$

Since the

$$\frac{4}{4}$$

is, according to the law of **inverses**, worth 1 and since, by the law of **identity**, multiplying by 1 leaves a value unchanged we are able to say that the true value of

$$\frac{28}{44} \text{ is } \frac{7}{11}$$

1.3.8 (6) The law of **multiplication by zero**.

Of all the laws this will seem to many people the most obvious. No matter how many lots of nothing one has it is all still nothing. Yet, in our modelling, we do often contrive to write models which include a multiplication by zero. Trying, within a program, to *divide* by zero is a mistake which is common enough to have caused a specific error message, 'division by zero' to be included in many compilers and interpreters. There is rarely such an error message for a *multiplication* by zero yet, when we are debugging programs, such mistakes, sometimes catastrophic, are not uncommon. This very obvious law is *unintentionally* broken all too often!

1.3.9 As a way of exploring the influence of the symbols and the laws in modelling let us return to our original example of the cows. If someone opens the gate to the field and allows three of them to wander off we might describe the situation by saying 'from our original group of ten we have allowed to escape (removed), three and this leaves a herd of seven'. Or, 'ten less three leaves seven' this is fully represented by the model $10 - 3 = 7$.

There are many possible variations on such arrangements of words. If, for instance, we did not see the cattle straying but merely knew that only seven remained in the field we might say, 'there were ten and now there are only seven; how many have strayed?'. In this case we have an unknown quantity which we need to find and our model might well be $10 - ? = 7$.

It is, in practice, very common for our models to contain unknown quantities which we need to find and we usually do two things to simplify the discovery process. First we put a stated symbol in place of any question marks and then we arrange our model so that *all that we know* is grouped together and *what we need to find out* is clearly identified. If, in the present example, we stipulate, LET (us agree that for this purpose that) '*S*' (will) represent the (unknown number of cows which have) strayed, then our model will be $10 - S = 7$.

Here, however, the 'knowns' and the 'unknown' are muddled together, but if we rephrase the description as, for instance, 'the number of strays is the difference between the original number of cows and the number still in the field,' we can rewrite our model as $S = 10 - 7$ and if we remember our 'subtraction facts' and we know that the difference between ten and seven is three, we can solve our problem.

It would not be worth the effort of writing a computer program to solve a single problem of this type since it would take more time than we should ever save. It is really only worth that effort if we use the computer to deal with many repetitions of solving precisely similar problems using the same model. If we were called upon to deal very frequently with simple 'difference' problems like this one we should devise a 'generalized' model where the numbers concerned could be input from the keyboard (or other input device), or from a stored file.

1.3.10 It is, nevertheless, very valuable to spend time thinking about and practising the building of such simple models because every mathematical statement has been developed from, and can be retranslated back into, descriptions in our ordinary language. Much of the technique and many of the methods for doing this is what we may have learned earlier under the name of 'algebra'. In fact that very word implies 'writing down complete statements'. In practice we simplify such statements by replacing words with symbols wherever possible. Many of us, however, do not feel too comfortable dealing with symbols only as we are, quite naturally, more at home with words. Mathematicians, on the other hand, know that words are actually very tricky things and may be interpreted very differently by different people. In order to avoid any possible ambiguity we use either internationally agreed symbols or we state what our symbols are to mean in any particular case. Symbolic statements can also be very brief and this too is a help to clarity.

Throughout this book we will discuss the meanings of the symbols we use, as we meet them, and try to make them as 'user-friendly' as possible.

1.4 Generalized models

1.4.1 There are five words which become important immediately we start talking about the next stage in modelling for computing: **assigning, constant, declaring, parameter and variable**. One problem for many of us when reading programming manuals is that we are not always comfortable with the language used and these particular words are common stumbling blocks. This is partly because at least some of them are rather different from what we meet in traditional mathematics. We will try always to discuss such differences of usage. (It is also perhaps confusing that in computing we make use of these same words when dealing with alphabetic, or 'string' – data. This is because all characters, alphabetic as well as numeric, are actually stored and manipulated as number codes so we need only consider them all as different data types. We, however, are only concerned with the true numerics and the mathematical symbols in this book.)

1.4.2 To examine the idea of a **constant** first.

If, for instance, we consider the comparison between the diameter and the circumference of a circle we find that the relationship is fixed, for all practical purposes. The *larger* the diameter the *larger* the circumference; not in any haphazard way but always with the circumference 'three and a bit' times as long as the diameter. Such a fixed relationship we describe as a 'constant', in this case labelled as 'pi' or 'π' (simply because they were the labels used in the original Greek explorations of that relationship and have been accepted as standard 'labels'). There is a special difficulty with π since we have as yet no fixed value for it so we may use a value for this constant which is appropriate to the task being performed. That value will be included in the data dictionary and be written into any program which uses it so that it stays the same throughout that application. We should also use a consistent value for it in any set of manual calculations.

1.4.3 Variables need both a name and a value and whilst the name, once decided will be fixed, the value may well change during operations on the model. Variables will *not* usually have universally agreed labels and each must be given one. Within the programming team, we must agree upon both the label and its interpretation. We must also be sure that each variable name is unique so that no ambiguity or confusion may arise. The chosen name will often be a single letter in traditional mathematics, but may be a complete word or phrase in modern programming languages. The label must be used consistently throughout the model. In our original example the word 'COWS' is the variable name. In programming we must **declare** all such labels at some stage, that is to say that we must include the labels in the program and the documentation. (In good programming practice we shall also record them in the data dictionary.)

It is then common practice to **assign** an actual value, or **parameter**, (meaning only a quantity or value) to that variable either within the program or by input from keyboard or file. This means that we shall **store, in memory, at the address allocated to the variable which we have declared, the value** which we intend it to have at any given stage of the program run. (Once again, in our example, the variable, 'COWS' is assigned the original parameter, 10).

If the program is to be used for a number of runs we must also, as a matter of good practice, initialize all variables to zero so that the processing is not corrupted by any random value stored at the relevant address or by a value remaining from an earlier run.

1.4.4 To illustrate all this let us make a generalized model of the 'missing cow' problem. First we label and declare the variables; LET (and this is a classical mathematical word for declaring variables), F equal the numbers of cows in the field to start with; (LET) R = the number of cows remaining and (LET) S = the number of strays.

The next step is either to **assign** or **input** parameters for each of the 'knowns'. Thus either:

LET $F = 10$; $F \leftarrow 10$ (if we are assigning within the program),

or

ACCEPT F (if we are taking keyboard input)

LET $R = 7$; $R \leftarrow 7$ or   ACCEPT R

(The left arrow, $\leftarrow$, is a common symbol for assignment; whatever is on the left of the arrow takes on the value of whatever is to its right.)

These steps together complete the process of establishing the 'knowns' and

LET $S = F - R$; $(S \leftarrow F - R)$; completes the processing

We now need only 'PRINT S' to convey the answer to us.

1.4.5 This is the essential core activity of all programs. **Operating as needed upon a model of the problem which the program is designed to deal with**.

1.4.6 From school mathematics we may remember a variety of models which we were required to learn. $A = L * B$, to find the area of a rectangle, is wholly made up of variables. $V = 4/3(\pi r^3)$, the volume of a sphere, on the other hand, has only one variable from which the unknown, V, is eventually derived; can you identify all the constants? As a useful task write down some of the models which you can remember and try to identify which elements are constants, which are unknowns and which variables. Then try rewriting the models in as many different ways as you can.

If we take $A = L * B$, for example, we can rewrite it as $L = A/B$ or $B = A/L$ so that, eventually, whichever two parameters *are* known we can derive the third, unknown. Try creating such variations of the $V = 4/3(\pi r^3)$ model.

1.4.7 One task which frequently falls to the programmer is to take an existing model and adapt it to accept some new arrangement of variables in order to solve the model for a different unknown. **Do not rely on remembering such models. It makes much more sense for the programmer to know where to find such them in the reference books**. It is also a waste of much energy to re-invent a model which someone has developed perhaps centuries earlier.

1.5 Spatial models

1.5.1 Many applications of computer processing are concerned with design, manufacture or movement in the real, three-dimensional world. To model real objects, real places and real journeys we need something more than simple numbers or symbols and we turn to pictorial, or graphical, representations and various geometries for our tools. The VDU screen and some other computer output devices are powerful means of *presenting* spatial information but our *processing* must depend upon numerical or logical representations of the three-dimensional world of space and time, since the ALU is limited to these two types of operation.

We have, traditionally, explored and represented such realities by technical drawings, graphs and maps. For the computer, however, whether the problem is one of automatically piloting an aircraft across oceans or causing the cutting edge of an automated machine tool to drill a cylinder block for a motor car, we have to make computer models which adapt or replace these traditional mathematical processes.

1.5.2 The primary tools for building such models are the **co-ordinate descriptions**, invented by the great French mathematician, Descartes. We may organize the data from such descriptions in the matrix or array and its simpler relation, the vector. Once again we shall find that, if we can construct the models, we shall be able to select from existing mathematical processes in order to perform the necessary operations upon them.

1.5.3 In this book we shall practise with two-dimensional problems only, but everything which we consider may be extended to three or more dimensions by comparatively simple expansions of the basic models and the techniques for their manipulations. It will often be helpful to see a sheet of graph paper itself as a model of the VDU screen and vice versa.

Cartesian coordinates set real values, both spatial and numerical, on to a square grid of which each axis was a number line, itself a model of our number system (as shown in Figure 1.1).

Figure 1.1   The number system as number line.

1.5.4 When two axes, at right angles to one another, are each designated by such a number scale and intersect at 0, problems involving two numeric variables or two-dimensional space may be modelled. Any unique position in the space denoted by the grid may be identified by stating two numbers, **the first** representing the distance of the point from zero **along the horizontal axis** and the **second** its similar distance **along (up or down) the vertical axis**. This can be seen in Figure 1.2.

1.5.5 Provided that **the order of the two numbers is always the same** each such 'ordered pair' describes a precisely identified, unique, point (see Figure 1.3 opposite).

Figure 1.2 The grid devised by Descartes.

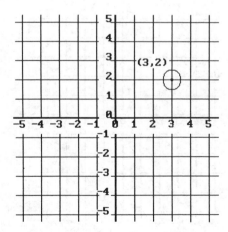

Figure 1.3 The point as an ordered pair.

Figure 1.4 $y = 2x + 1$ as a graph.

Two or more such points may define a line. You are probably already familiar with the description of the relationships between two variables by the line on a graph plotted from such points, as illustrated in Figure 1.4.

This also gives us exactly what we need for our spatial modelling, a simple means of converting a position, or positions, in space into a purely numerical description.

1.5.6 When we move from one such point to another we can simplify that journey, in purely spatial terms, into a difference between our first position and that at which we finish. An example is given in Figure 1.5.

Figure 1.5 The journey from (1,1) to (5,3) = (4,2).

On plotting the two positions on our grid the *difference* between the first and last horizontal distances will give the first number and the difference along the vertical axis the second, in another ordered pair; this time denoting movement rather than only position. **Such a number pair is described as a vector** although the term also has the much broader meaning of any line of which we know both the length and the direction.

When we use our grid system to describe shapes or their movement we may well identify a number of points, a triangle, for instance needs three, as can be seen in Figure 1.6.

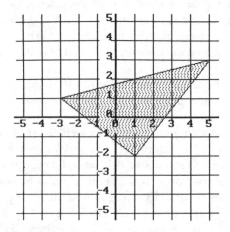

Figure 1.6   A triangle as ordered pairs.

1.5.7 Each of such points will have its own ordered pair. These ordered pairs will themselves need to be put into and maintained in a fixed sequence and an existing mathematical model, which has ready-made rules and operators for processing, is the matrix or array. As a matrix, purely numerical model, our triangle may be stated as:

$$\begin{bmatrix} -3 & 1 & 5 \\ 1 & -2 & 3 \end{bmatrix}$$

The mathematics of matrices existed before modern computers were invented but they might well have been made for each other as the machine is so well adapted to the manipulation and processing of matrices. Under the name of arrays they have evolved into one of our most flexible and powerful data structures.

1.6 Logical models

1.6.1 One part of mathematics which is often taught very badly is the application of logic. Even students who have been taught 'sets' rarely understand the relevance or the rigour of what they have learned yet, of all the fundamentals of mathematics, this study ranks as being of the very greatest importance in computing. We shall examine this topic in greater detail later but let us try to establish the basic principles.

It is an age-old principal that **we can only logically compare like with like**; so that it is, for instance, quite ludicrous to make statements such as 'She is more beautiful than he is tall'. It would be equally ludicrous, but even more of a trap, to say 'on a scale of one to ten for beauty she rates eight whilst he, on a similar scale for tallness, rates only six'. Can you see and explain why? One of the great problems of mathematical thinking is to decide just what may properly be described by a measurement and what is, and always will be, just a matter of opinion.

That division is between **parametric** (attributes which can be measured or given *parameters*) and **non-parametric** mathematics and modelling.

1.6.2 It is also of fundamental importance to realize that any measurement of one characteristic very rarely tells us anything whatever about any other characteristics of the items which we are trying to compare. To say that one person weighs more than another tells us nothing more than that; it certainly does not justify, for example, inferring or implying that the heavier person is taller, or more cheerful. For this kind of modelling we have to learn to think much more clearly and in a more disciplined way than that.

1.6.3 To start with we shall need to consider logical relationships between objects, numbers, statements, people and everything in the universe which may be logically compared. Instead of writing out a long list of such possibilities each time we use the word '**entities**' to stand for any single item of whatever kind it may be. An entity, then, is anything which has a logical existence of its own. When an entity is included within a set it is described as an **element** of that set.

Entities, in turn, may have many different features, qualities, measurements, characteristics and so on; different aspects of the entity which may concern us. The word we use to avoid writing the list every time is '**attribute**'. The final word, nowadays in quite common use in this context is '**set**'. It is often thought that any collection of entities is a set but this is not so. It is of the greatest importance, before embarking on any study of logical relationships, to be quite clear as to what constitutes a *mathematical* set.

1.6.4 The rules are brief but very clear:

❑ **A set must be defined**. In this case the word has a much more rigid meaning than usual. The set, whatever our purpose in using it, must be so clearly defined that we can say, of any entity, that it belongs either IN or OUT of the set; there *cannot be any 'maybe's'*.

To take an example, we cannot have a *mathematical set* of 'people with blue eyes' since there would be endless arguments as to whether one person's eyes were blue or grey or another's were blue or green. **If it is possible to argue whether an individual entity should belong in a 'set' or not it is not a true set.** This is the real meaning of **'two-state logic'**, which is all the computer can truly deal with in ordinary practice. Is there a set of tall people? You should, with a partner, 'define' some sets and then test them to see whether they truly conform to this rule of definition. Do this carefully and test your definitions against some extreme cases. This is how we create the test data which will truly expose some of the logical errors which cause our programs not to run successfully or reliably.

❑ **Every element in a set must be unique.** There are no duplicates in true sets. This immediately rules out many of the sets we see in everyday life. It may be entirely sensible, in ordinary life, to talk of 'a tea-set', for example. It cannot, however, be a *logical* set since it contains, say, six identical cups, six identical saucers and six identical tea-plates. In the equivalent logical set this would be reduced to just one of each of the items. Furthermore if we put together, for example, the set of numbers, 2, 3, 4 and 5 with the set of numbers 4, 5, 6 and 7 we would have the set 2, 3, 4, 5, 6 and 7. This is one of the situations where 4 + 4 does not equal 8 since we are adding two sets of four items together but have a set of only six items at the finish. Once again you should try a number of such 'additions' to be sure that you understand what is happening.

We may already be quite familiar with these ideas. In retrieving and sorting records in our computer files the allocation of unique 'key' fields plays a vital part. The combining of sets is what happens when we merge two files. What would happen during a merge if two records in the originally separate files had the same number in the key field?

Strictly speaking any filing system, including traditional manual ones, must abide by these rules if we are not to lose track of items in those files. Many program failures stem from inadequate understanding of the importance of uniqueness.

1.6.5 One common misconception is that all the elements of a set must have an obvious attribute in common. This is not a true test since we may group entities into any set which suits our purpose, so long as it obeys the rules. If

I wished to do so I could define the 'set of things in my left trouser-pocket'. It would, as I write this, consist of a bunch of keys, a handkerchief, a pocket-knife and a small wallet. The only attribute those things have in common is that they are, at the moment, in the relevant pocket! Such sets may not be especially useful in routine data processing but they are entirely valid nonetheless.

1.6.6 As other logics are developed the situation is changing slowly. In applications such as 'artificial intelligence' for example, we make use of what is called 'fuzzy' logic, a system where the two-state logic is replaced by a less rigorous one where 'nearly' or 'maybe' supplement 'is/is not'. But for all conventional information processing our databases, files, records and fields must conform to these two-state conditions.

1.6.7 Few interpreters or compilers can deal directly with the notation or the manipulation of sets. It is we, the human beings, who have to learn to think in this way, to allocate material to such logical structures and to plan and program the manipulation of information in such structures. The one exception is with a database management system (DBMS) which operates strictly on logical sets.

Even here, however, it is we who have to make sure that the two-state logical interpretation of the information is properly established.

As with all mathematical modelling there are consistent symbols and notations to be mastered if we are to be truly fluent in interpreting such structures.

1.6.8 One of the most practical structures is the Venn diagram (shown in Figure 1.7), by which we can show quite complex relationships but which develops from the relatively simple notion that we can define a set by 'fencing it in' within a circle (or an ellipse), so that any entity outside the circle is a 'not' and anything within is an 'is'. We can then combine circles to express very complicated relationships within and between sets. Any entity may belong to many different sets. Try making a list of the sets to which *you* belong.

Test them very carefully for definition but you may find that the 'uniqueness' criterion is less of a problem. Why is this?

In order to limit the complication of what we are trying to organize into sets, and any 'family tree' of sets is known as a **taxonomy**, we first **define the whole area of our problem**. This was known, in more formal times, as the 'universe of discourse', the actual and limited world we are examining or discussing. This has come to be described, a little more simply, as the '**universal set**' and in Venn diagrams is usually denoted by a rectangle surrounding all the sets in which we are interested. A company's stock list, for example, may be seen as a 'universal set' which will contain many other sets.

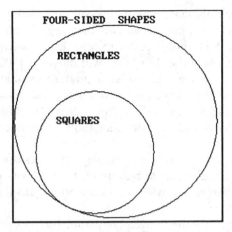

Figure 1.7 A universal set and two of its 'inner' sets. Can you interpret the story they tell?

Finally, a set which had no relationship with any other set would be a very strange one indeed. It is the complexity of how sets relate to one another and how these relationships may be expressed in the simplest possible way as 1s and 0s which we shall need to study in detail.

1.7 Statistical models

1.7.1 There are many situations where 'ordinary' numerical mathematics cannot provide all the models or tools which we need.

In very complicated situations, such as describing a whole country, actually counting or measuring *everything* about that situation would take too long, cost too much, need too many people or pose other very complicated measuring or counting problems. We sometimes describe such situations as **complex states** and we can usually deal with them only by taking samples and using those to give us insights into the whole complex. A computer system is actually just such a complex state because of the vast numbers of operations, of data items handled and stored and the sheer speed at which it operates and we are dependent upon statistical models and methods to plan and monitor our systems, their loading, reliability and performance.

1.7.2 The second area of interest which can only be modelled statistically is the situation where complete testing destroys what is to be tested. Manufacturers of light-bulbs, for example, are often required to certify and guarantee the life expectancy of their products. No-one would buy such things from a manufacturer who could not assure us of a reasonable period of reliability yet the only way to know for certain how long a light-bulb will last is to keep switching it on and off and 'running it' until it fails.

It would, however, be quite absurd to offer to sell such a lamp with the guarantee that 'We know that this lamp will burn for 1500 hours because it has already done so!' We expect reasonable guarantees of life expectancy and reliability from most things we buy from electric razors to motor cars and aeroplanes; especially aeroplanes! Yet here actual measurements of tests to failure are useless to us and only statistics allow us to model the problems.

1.7.3 Yet another area is where, based on the recorded history of past events, we need to be able to make reasonable forecasts of the future. Weather forecasting is just one very obvious example and again the modelling tools are statistical.

In weather forecasting the models are astonishingly complicated and the corresponding computer programs extremely complex. In these and similar areas we are much concerned with 'trends' or 'tendencies' and trend-analysis is a powerful management tool.

1.7.4 Finally, although we shall not delve into non-parametric statistics in this book, there is the analysis of opinion. The marketing of products is just one field where statistical models form a major source of management information based on the evaluation of opinion. An example of a statistical model is given in Figure 1.8.

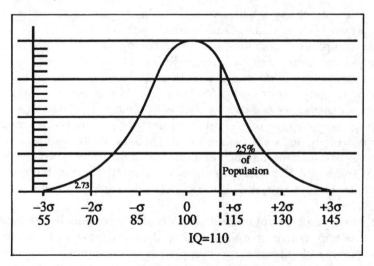

Figure 1.8   One of the commonest statistical models. Many human characteristics are distributed in this way.

1.7.5 What all these aspects of the real world have in common is that in all of them, by their very nature, we are denied the certainty of precise measurement. What our statistical models and processes allow us to do is **To diminish the extent of our *uncertainty*.** One of the saddest spectacles of modern applications of mathematics is that so often the 'persuaders' ignore that this is the fundamental nature of statistics. They try instead to delude us by saying '... statistics *prove* that...' this is a manifest untruth. **Statistics deals in tendencies, likelihoods and similar *un*certainties** and whilst it may offer us evidence that something is more or less likely to happen, or to be the case, it can never *prove* anything much less tell us *why*.

1.8 Algorithms and trial-and-error

1.8.1 Traditionally mathematicians have always sought to create 'algorithms' for processing their numbers and the word meant 'a fixed procedure by which, if we provide the right inputs, follow the procedure correctly and obey all the mathematical laws whilst doing the correct mental arithmetic, we will find right output'.

The 'drill' by which you were taught to perform 'long' multiplication is just such an algorithm. Algorithms relieved us of the need to re-invent processes every time we needed them. But we had to memorize them and make sure that we didn't forget any steps or how to perform them; a tall order for many of us.

A properly designed and written computer program, which has been correctly tested, should be an algorithm and provide reliable and effective processing so that, for the user, **right inputs = right outputs.**

1.8.2 However, we have, as an additional tool, the speed of the computer and this can be put to good use in running 'try-it-and-see' models. The formal name for these is **heuristic** models and they allow us often to avoid constructing an algorithmic model in favour of one in which we try out various combinations of values in order to see if we can find a suitable solution to a problem. Spreadsheet systems often allow very sophisticated 'what if?' models and we should understand the nature and application of these to everyday problems.

1.8.3 We must also bear in mind that in real life it is also often the case that it is much more valuable to have a usable answer quickly, perhaps a sensible approximation, rather than a long-delayed 'precise' answer.

At this point it is wise to consider more carefully a symbol which we undoubtedly take too much for granted. The equals symbol (=) actually indicates a rather limited relationship. It does not say, for example, that what is on its left is *exactly the same in all respects* as that which is to its right. The expression 3 * 2 = 6 merely says that 3 * 2 is *equivalent*, that is, it has the same *value* as 6. If we wish to say that two entities are exactly alike in all *mathematical* respects the symbol which indicates this special condition is written as '≡' and we describe the relationship as **congruence**. In real life, however, and in our computer models of reality we very commonly deal with situations where, 'approximately the same as', 'not more than', 'at least as much as' and similar conditions are relevant, rather than equivalence or congruence. The equals sign has been taken for granted during so much of our schooling that we may find it difficult to see intelligent approximations as being just as important, as indeed are the *non-equivalence relations* or *inequalities*.

There are several symbols in use for 'approximations' but in this book we shall use (≈) throughout for the sake of consistency. Similarly you may come across different ways for writing the 'not more than', (≤), and the 'at least', (≥), but their use in the book will be consistent. The symbols for 'more than' and 'less than', > and <, respectively, are used more consistently and should present no problems.

1.8.4 To see our inequalities a little more clearly let us look again at the graph of $y = 2x + 1$. The line itself consists of all the *points* on the graph where the statement is true.

That line is, then, the line of the *equivalence*. What of all the rest of the surface indicated by the grid? Take any point below and to the right of the line, say (2,2). Here the $x = 2$ but so does the y and since the y only *equals* the x it must be *less than* $2x + 1$; or, $y < 2x + 1$. If we try a few more such points, even those very close to the line of $y = 2x + 1$, we shall find that this whole region of the graph is described by $y < 2x + 1$ (see Figure 1.9).

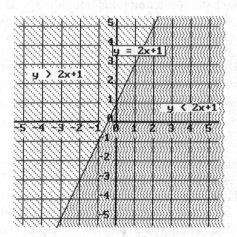

Figure 1.9   The graph of $y = 2x + 1$.

1.8.5 What about taking that region and *including the line*? Can you write down the model? Next look at the region *above and to the left of the line*. What will be the **inequality model** which describes that region? What will be the model if we include the line?

Can you see that every time we state an equivalence relationship we are also implying a whole set of inequalities? Test out some others of your own devising and try to get used to seeing a graph as a whole set of relationships as well as the one which we choose to 'pick out'.

Here again is a small but extremely important section of mathematical ideas where the testing of our programs is concerned. We must decide, whenever we program any equivalences, just how our program should deal with the corresponding *inequalities* and what values our test data should include in order to reveal any defects or omissions in design which have arisen from ignoring them!

1.9 Modelling in pseudo-code

1.9.1 Whatever *mathematical model* we select, it is still only **man-readable**. The stage must ultimately be reached when we must make it **machine-readable** by designing a program segment to translate it into computer-related form.

1.9.2 Unfortunately each high-level computer programming language has its own ways of doing things and many even have numerous 'dialects'. There are, for instance, many variations of BASIC, but just as a good program should, ideally, be 'ego-free', that is to say in a standard style which is independent of the whims of the individual programmer, so should a good program *design* be 'language-independent'.

1.9.3 This means that the design should be practicable whether it will eventually be programmed in BASIC, for example, or C++. It must be possible to test the logic and the planned operation of the program *before* it is actually coded in the language of choice.

It also means that the designer, after the design and testing processes, can choose which of a range of languages is best suited to the particular application. If, however, the program is designed from the beginning in the form and terms of one language it may be very difficult and costly to *redesign it* for eventual application in another.

1.9.4 The chosen solution in times past was the **flow-chart** and this still has value. It is, nevertheless, a great advantage to be able to design a more directly program-related model which can be written in simple English, or any other everyday language, using standard programming **constructs** and we can do this in **pseudo-code**.

1.9.5 The essential constructs are:

(1) Sequence and in pseudo-code form this is:

Process A
Process B
Process C
END

with each part of a complete program run being completed in turn.

1.9.6 (2) Selection of which there are several forms. Among them we may identify several varieties of a skip, or branching, to another process or routine.

(2a) Selection by single if: pseudo-code:

IF condition
    DO process
ENDIF

or

WHILE condition
    DO process
WEND

(2b) Alternative if: pseudo-code:

IF condition
    THEN DO process A
    ELSE DO PROCESS B
ENDIF

(2c) Multiple selections: pseudo-code

IF condition 1
    THEN DO process A
        ELSEIF condition 2
            THEN DO condition B
        ELSEIF condition 3
            THEN DO condition C
    ⋮

ENDIF

which may continue for any number of conditions subject only to any limitations in the language in which the program is eventually to be written.

1.9.7 (3) There is also the loop, which may be **unconditional** or **conditional**, single or **nested**, and which may be **recursive**.

(3a) A single loop may be, in pseudo-code:

```
read N
   DO K = 1 to N STEP 2
      process
   ENDDO
END
```

in simple, 'loop-counter' form.

(3b) DOWHILE condition

```
DOWHILE condition
   process
ENDDO
```

controlled by a 'logical' evaluation of 'condition'.

(3c) DO UNTIL condition

```
DO UNTIL condition
   process
ENDDO
```

again a logical decision to end the loop.

(3d) The **nested loop** pseudo-code:

```
read N, N2
   DO i = 1 to N STEP 1
      DO j = 1 to N2 STEP 1
         process A
      ENDDO
   process B
   ENDDO
END
```

may be nested to six levels 'deep' in many languages.

There are some conventions and words associated with this form of modelling among which is the **indentation** of constructs within constructs as an aid to easier reading.

1.9.8 The use of words is fairly simple because of the comparatively plain English coding. We tend to use:

❑ **READ** to imply that data is retrieved from a file
❑ **ACCEPT** to indicate a pause for data input from keyboard
❑ **DISPLAY** to show output on a VDU screen
❑ **WRITE** to output data to a file and
❑ **PRINT** to send output to a printer.

In addition to these, for working examples or setting up test data we may use DATA to indicate that what follows is a data list to be READ into the program. In other words we may emulate the contents of a file.

1.9.9 If we put some of these ideas into a specimen program segment we may try our hands at identifying what happens at each line and do a pencil-and-paper test run with some different data.

```
read NAME, AGE
   DOUNTIL end of file
      IF AGE <30
         display NAME
         accept DEPARTMENT
         print NAME, AGE, DEPARTMENT
         write NAME, AGE, DEPARTMENT
      ENDIF
   read NAME, AGE from next record
   ENDDO
END
```

1.10 Conclusion

The choice of appropriate models will usually be a matter of selecting from what is already available in the mathematical sources. Whether such models will fully serve our purposes or will need modifying or whether new models will need constructing from scratch is a matter for careful analysis and judgement. It is we, in consultation with the users, who need to have the understanding and the knowledge to make such judgements and to select or develop models accordingly. Remember that many of those which we need will have already been developed and tested and these, if we know what to look for and where, will save us a great deal of time and hard work.

Exercises

(1) Give a numerical model which represents the solution of:

5 cows + 3 water buffalo + 3 goats.

(2) What is the result of $5((10 + 7) * (3 + 4))$?

(3) Which of the laws in 1.3.1 *do not apply* to subtraction and division? Show why.

(4) $\sqrt[3]{R(3/4 V)} = \pi$: what does this model represent when written with V as the subject?

(5) What variable types and parameter ranges would be appropriate to:
  (a) a person's age
  (b) a person's sex
  (c) a date
  (d) a time
  (e) the price of a motor car
  (f) an hourly wage rate
  (g) the number of hours worked by an employee in a week?

(6) Write down two ordered pairs from the graph in Fig. 1.4 where the values for x and y are whole numbers.

(7) What ordered pair represents a journey from (3, 5) to (–5, –3)?

(8) Name three types of situation where statistical modelling is essential.

(9) Write a list of: **(a)** six *proper* sets and **(b)** six groups of entities which are not proper sets. Give reasons in both cases.

(10) List at least ten sets of which you are a member in the form: $I \in \{...\}$.

(11) Identify ten real-life situations, five where values must be 'at least ...' and five 'not more than ...'.

(12) Write a pseudo-code model of a procedure to accept and validate a user-code and password to permit or deny access to a computer system.

2

Constructing models

Objectives

After completing this chapter you should be able to:

❑ develop a model following an organized process
❑ convert a specified situation into a model
❑ construct models of simple commercial and other real situations
❑ understand the differences between algorithmic and heuristic models
❑ make models using matrices, tables, trees and diagrams.

2.1 The modelling process

2.1.1 The process of actually developing or constructing a model can be organized into a number of logical stages. If we then follow those steps in a deliberate sequence we are likely to find the process becoming a good deal easier. Like everything else, it needs practice to develop confidence.

2.1.2 The development of the model may be reduced to a sequence of stages (we shall often refer to this sequence) given below.

Stage 1. Is the problem truly one which demands a mathematical solution? If the problem is numerical or strictly logical the answer will undoubtedly be 'Yes'. If the customary response to such problems is simply an *opinion* the answer, in all probability, will be 'No'.

Stage 2. What *relevant* facts are known? (When we first encounter a problem the biggest difficulty, often, is that there is just *too much* 'information', which often confuses the issue.)

Stage 3. What other facts do we need and where can we find them?

Stage 4. Is there a *natural,* or a desirable, sequence of steps in dealing with the problem?

Stage 5. Can the facts be represented by symbols and separated into 'knowns' and 'unknowns'?

Stage 6. Is there an *existing* model which will serve, or may be adapted, in order to save us the labour of inventing a new one?

Stage 7. Is the existing model one which makes good use of the capabilities of the computer?

Stage 8. How should I express my model in terms which the programmer can understand?

2.1.3 When we consider the problem we must first apply a little common sense. If, for example, someone asks the question *'What is the weather like?'* we should not immediately rush to create a mathematical model and a computer program in order to find the answer(s).

First we should ask 'Who wants to know and for what purpose is the information needed?'. If it is a member of the family who only wants to know whether to take a raincoat and umbrella, an opinion, based on just looking out of the window, may be all that is expected. If, on the other hand, the question is being asked by the pilot of an airliner due to fly 400 people to New York the only useful answer will involve the gathering, processing and presentation of a great many facts. In the first case a computer is neither needed nor justified; in the second case the model will be highly sophisticated and both the computer and the program(s) will be very complex.

Such examples often seem only too obvious but people do, frequently, plunge straight into doing all sorts of complicated sums and programming without pausing to ask whether it is necessary. True mathematics is not like this at all; indeed it has been said that it should be '90 per cent think and 10 per cent sums'. Always pause and *think* about the real nature of the problem *and* consult the user.

2.1.4 Let us follow the process of simple modelling through the stages with an everyday example:

The management of an engineering firm which produces aircraft parts has agreed to pay its skilled workers at an hourly rate for a basic week of up to 40 hours with any time an employee works over 40 hours paid at one-and-a-half times the normal rate. Employees are only paid for completed quarter hours. The firm does not normally work during public holidays, but any hours which *do* need to be worked then will be in addition to the normal 40 hours and will be paid for at twice the normal rate. All employees pay 6 per cent of their wages into a pension scheme. A payroll system is needed.

Creating such a 'narrative' description of the problem, i.e. a 'story' which is agreed to by the user as a correct representation, is a vital first step. However. words are often treacherous and also any verbal description will almost always include much which, although relevant to the situation and perhaps to the full final solution, is not essential to the initial model. It is here that the process of continued simplification begins.

The problem is how is each employee's wage to be decided?

(At each stage you should discuss other possible ways of looking at the problem or other solutions which you or your fellow students can suggest.)

2.1.5 The stages.

Stage 1. Yes, it is numerical since it deals with measurements of time and money.

Stage 2. Relevant facts:
(i)   up to 40 hours at basic rate
(ii) anything over 40 hours at 1.5 times basic rate
(iii) work on public holidays paid at twice the basic
      rate
(iv) final pay will have 6 per cent deducted and paid into pension fund.

Everything else in the original 'narrative' description of the situation is now redundant so far as developing the model is concerned. Look back and see how much of it could be crossed out.

Stage 3. What we shall need to know, for each employee, will be:
(i)   basic pay rate (fixed, except for possible periodic reviews)
(ii) total hours worked in any given week
(iii) any hours worked on public holidays in that week.

This identifies **a data-acquisition task** which we shall, eventually, have to build into the system but for the present we can, having noted the need, continue developing the model.

Stage 4. The 'natural' stages would seem to be to calculate:
(i)   basic pay (there may be nothing more)
(ia) public holiday pay, if any;
(ib) 'overtime', if any;
(ii) pension fund deduction.

Stage 5. Symbols, constants, variables. Stages 3 and 4, as they are written above, are very 'wordy', so we can be pretty sure that variable names and symbols *will* simplify the task. Let us first 'declare' our variables and constants.

In the following list we can, in our minds, substitute for the word 'LET', the words, 'Let us agree that throughout our model...'. We can also interpret the 'equals' sign as 'stands for' or 'symbolizes'.

LET R = the hourly pay rate
LET T = the total hours worked
LET V = the ordinary overtime hours
(why do I choose *not* to use 'O' for 'overtime'?)
LET H = the hours worked on public holidays
LET W = the total amount of wages to be paid to the employee
LET P = the amount to be deducted and credited to the pension fund.

This does not decide the *types* of data item and we must think about that a bit further. We will assume that R implies pounds and pence, or dollars and cents and this poses problems. We should normally use **decimal numbers** but these **often result in errors when converted to binary in the machine** whereas whole numbers do not. Since overall amounts in this case are comparatively small it would be wise to use whole numbers and avoid the fractions. We can do this by stating R in pence (or cents). This means declaring R as an INTEGER and, since it will be fixed, at least for quite long periods of time, we may well treat it as a CONSTANT.

T, V and H, on the other hand, *will* involve fractions but only 0.25, 0.5 and 0.75. (Can you see why?) These can all be converted to bicimal fractions in the machine without loss of accuracy so we can declare these VARIABLES (since they will change from week to week) as REAL (numbers), or REALS.

We will investigate these number types more thoroughly in a later chapter but for the moment we need only accept that integers *cannot* be other than whole numbers whereas REALS may have a fraction part as well as an integer part; $2\frac{1}{2}$, 2.5, 3.63 and so on, are real numbers.

The final pay, W, will need to be set out as dollars and cents (or pounds and pence) at the end of processing but can be calculated throughout in INTEGERS and only converted to REAL at the last possible moment in order to keep error to a minimum. There is, similarly, no reason why the pension contribution, P, should not be calculated as integer and converted to real at the last possible moment. Almost always there is another 'bonus' to using integers. Most computers show an appreciable gain in speed when processing integers. Do you know why?

You should always discuss the choices of data types with your colleagues and make sure that you understand what governs our choice. See also if you agree with the suggestions I have made.

2.1.6 It is at this stage that we should, in practice, begin to compile the data dictionary in which, as part of the program documentation, we record data assignments, types and structures. We should also decide on and include in the dictionary the permissible RANGE for each data item. For example, the range for *T must* be 0–168. Can you see why it cannot be less than 0 or more than 168 or why *V* and *H* should be not more than 128? Try to establish all the necessary data ranges because eventually we shall need them in order to stipulate test data, among other things. (Why should we test *T* for 'less than' 0 and 'more than' 168? One of the advantages to working as a team in computing is that we can discuss such points with our fellow members and it is important that you should get into this habit as you practise model-building.)

2.1.7 We are now ready to attempt to build the model itself and there may be many different ways of doing this. We may not arrive at the 'best' model at the first try. First and foremost we need a model which *works* (is effective), continues to work *reliably* and is easy for anyone else to understand and to *maintain*. We may describe such models as 'robust'.

Once we have such a model we may, if it proves slow or inefficient, (or if we want to make it more user-friendly, for example), 'tweak' the model into a somewhat different form. If we make a mistake we can always go back to an original *which we know to be robust*.

For our present purpose a first step might be to state that

IF $T \leq 40$ THEN $W = (R * T)/100$

The '/100' element of the model will convert cents back to dollars and cents. This model is only part of what is needed since it ignores the pension plan deduction of 6 per cent so

IF $T \leq 40$ THEN $W = 94\%$ of $(R * T)/100$
$\qquad$ (or) $W = (((R * T)*(94/100))/100)$ and
$\qquad\qquad P = (((R * T)*(6/100))/100)$.

(When we are modelling it is quite sensible to use brackets to show each step, to 'nest them' if necessary and finally to check that the number of left brackets corresponds to the number of right brackets. Most computer translators will signal a run-time error if we muddle the brackets but we can't be entirely sure – so check carefully and practise!)

2.1.8 Before we do anything else let us test this crude model, bearing in mind that we only do one step at a time and that the brackets show us the sequence in which we do them. If multiple brackets are 'nested', that is to say with pairs within pairs, we calculate the contents of the innermost pair first.

So, if we ASSIGN 'invented' values of 30.5 to '*T*' and 250 (cents) to '*R*' we have

$W = (((250 * 30.5)*(94/100))/100)$

and only the arithmetic remains to be done, in four further steps

$W = (((7625 * 94)/100)/100)$

$W = ((716750/100)/100)$

$W = (7167.50/100)$ cents

$W = \$71.675$

Now, by the same process, check that $P = \$4.575$ by substituting our test values of 250 and 30.5 for R and T in:

$P = (((R * T) * (6/100))/100)$.

If this simplified model works we can be fairly sure that we are on the right lines, with the additional incentive that eventually all the drudgery of the arithmetic will be performed by the machine!

2.1.9 So far, however, we have looked at the basic week of not more than 40 hours. If the week worked was more than 40 hours a different strategy must come into play

IF $T > 40$ THEN $V = (T - 40) - H$

and we can fit this into the overall model

$W = 94\%$ of $((40 * 250)+(V * 1\frac{1}{2} * 250)+(H * 2 * 250))$ cents

which is sound, but perhaps a little clumsy. If we now apply the distributive law we can start to simplify things a little

$W = 94\%$ of $(250 * (40 + (V * 1\frac{1}{2}) + (H * 2)))$ cents.

By substituting $((T - 40) - H)$ for V (as we worked out, above)

$W = 94\%$ of $(250 * (40 + (((T - 40) - H) * 1\frac{1}{2}) + (H * 2)))$ cents

$W = \$(94 * 250 * (40 + (1\frac{1}{2} * ((T - 40) - H))+(2 * H)))/100 * 100$ and

$P = \$(6 * 250 * (40 + (1\frac{1}{2} * ((T - 40) - H))+(2 * H)))/100 * 100$.

Check the model by choosing test values for T, and H and working through the steps of calculating the contents of each pair of brackets in turn.

2.1.10 If we had been solving a conventional mathematical problem which asked us to 'simplify' the equation, these models would still be capable of a good deal of refinement and in an examination we should no doubt lose marks if we left our work at this stage. Given the speed of the computer, however, it is better *not* to pursue matters further since the model is 'workable' and would be reasonably clear to a colleague as all the parts are still clearly identifiable.

2.1.11 What remains to be done is to connect our four partial models into a sequence so that:

IF $T \leq 40$ THEN $W = \$(((R * T)*(94/100))/100)$ and
  $P = \$(((R * T)*(6/100))/100)$
ELSE $W = \$(94 * 250 * (40 + (1\frac{1}{2} * ((T - 40) - H)) + (2 * H)))/100 * 100$ and
  $P = \$(6 * 250 * (40 + (1\frac{1}{2} * ((T - 40) - H)) + (2 * H)))/100 * 100$.

Stage 6. (and 7). This is certainly not a set of conventional mathematics textbook models, like $A = L * B$.

2.1.12 Where else might we look? The answer, of course, would be to consider the use of a commercial payroll package in which such models would certainly be found. If, however, the user could not justify the cost of such a package we might still be asked to produce a simple program. What else is available to us?

Could a spreadsheet do the job? We should, in that case, still need our model and we might do the rest by writing suitable macros within the spreadsheet.

Stage 8. If we need to write a program the final stage will be to express our model in pseudo-code form or a program structure diagram. Throughout this book we shall use pseudo-code since it is closest to a plain-language description of the task and is easily converted to any chosen high-level language when coding begins.

2.1.13 Since we need to use our program for a number of employees we shall eventually need to embed our 'core' program segment in a loop which will deal with *all* employees but, for the moment, we need only embody our models in a simple SELECTION routine

```
H = 0; P = 0; R = 250; T = 0; W = 0;
read H, T
  IF T ≤ 40
    W = (((R * T)*(94/100))/100)
    P = (((R * T)*(6/100))/100)
```

    ELSE
$W = (94 * 250 * (40 + (1\frac{1}{2} * ((T - 40) - H)) + (2 * H)))/100 * 100$
$P = (6 * 250 * (40 + (1\frac{1}{2} * ((T - 40) - H)) + (2 * H)))/100 * 100.$
    ENDIF
write *W,P*
END

2.2 The algorithmic model

2.2.1 This step-by-step process has brought us to *one* possible solution to the problem, there will be many possible variations and you should try some other ways of modelling this task.

What we have arrived at is a straightforward algorithm. If we input the right values for *T* and *H* (and *R* is kept up-to-date) we must arrive at the right answers.

2.2.2 An example of a similar simple commercial model which we might meet is

A finance company charges 15 per cent per year on all loans. A program is needed to calculate the total amount to be repaid after any given number of years.

We can, this time, abbreviate some of our steps.

Stage 1. Yes.

Stage 2. Fifteen per cent per year.

Stage 3. Amount borrowed and number of years the loan 'runs'.

Stage 4. Yes, a new calculation for each year.

Stage 5. *A* = amount borrowed (variable); *D* = the amount due to be repaid (variable); *R* = rate of interest as a percentage (constant); *T* = the time for which the money has been borrowed, in years (variable). (Discuss the best numeric data *types* to use.)

Stage 6. Yes – substituting our chosen symbols for the traditional mathematical ones in the conventional 'compound interest' formula $D = A(1 + R)^T$.

Stage 7. Probably not. Most people would not immediately recognize what this formula does and that is a very good reason for not using it in a program.

It will be simpler, clearer and easier to test and maintain if we set up a more straightforward model which will repeat the calculation, by a loop, for each year. The sheer speed of the computer makes such repetitions a practicable alternative to the one-time algorithm. Let us try to create such an 'iterative' (i.e. repetitive) model.

At the end of the first year the borrower will owe

The amount borrowed (A) + 15% of that amount $(R * A) = D$

At the end of the second year the borrower will owe

D + 15% of $D = D$ (the new value of D)

and so on, for the number of years the loan runs.

When we use a value arrived at in the previous loop in order to calculate the current 'pass' we call this a RECURSIVE iteration (and a very powerful tool it can be).

In the present case we have a fairly simple model

$$D = D + ((15 * D) / 100)$$

and if we start with inputting values for the two variables, A and T we have a complete model.

2.2.3 Stage 8. Pseudo-code:

```
A = 0; D = 0; R = 15; T = 0.
read A, T
D ← A
   DO UNTIL T = 0
   D ← D+((R * D)/100)
   T ← T – 1
   END DO
write A, D
END
```

This model is capable of being developed for many commercial tasks. It could, for example, be converted to monthly calculations for, say, a credit card company. Once again you should experiment with the model, adapt it to other purposes, test it by 'dry running' and so on.

2.2.4 Discuss other simple commercial or industrial problems with your fellow students and systematically work through the process of modelling them. Don't worry if you and your colleagues come to different conclusions, there will almost always be a number of possible solutions to any problem. If it works, when you dry-run it, your solution is likely to be practicable. It is also important to remember that almost any solution is capable of refinement and improvement; the real test is whether your model is well thought-out and reliable.

Do not ever be content simply to be told 'This is the answer'. You may well develop better, more efficient answers by using your own experience and your own ideas.

2.3 A model of a game

2.3.1 Let us, in order to build up some versatility, look at a different kind of problem.

A popular English table game is known as snakes and ladders. On a very brightly coloured board, of 100 squares numbered in sequence, are printed several snakes, and a number, not necessarily the same, of ladders, of different lengths (see Figure 2.1). The players move their counters around the board in turn, according to the roll of the dice. As each moves along the squares he or she may land on a square containing the bottom of a ladder or the head of a snake. In the former case the player moves to the square at the top of the ladder and *gains* ground: in the latter case the player is 'swallowed' by the snake and slides right down to its tail. The first player to reach the 100 square wins the game. It is a simple game of chance but can be great fun with plenty of 'highs' and 'lows' along the way.

Figure 2.1 A black and white, simplified, snakes and ladders board.

2.3.2 Many people, faced with writing a computer program for a snakes and ladders game, would be pre-occupied with the graphics. This is *not* the best starting point! It is especially important that we should stop and think! We must get the action right and then, much later in the programming sequence, we may wish to restore full-colour graphics to the screen!

As always, let us simplify the model. This means eliminating all *unnecessary* detail (so far as the model is concerned).

First of all the snakes and ladders may be represented by simple arrows and we may substitute lines going from one position to another (or vectors), at this stage. See Figure 2.2.

Figure 2.2   The simplified board (part).

The board, although *presented* as a square, because in practice it may be folded to fit an oblong box, is not truly a square but a strip!

If we follow the movement of the counters, shown by the curved arrows at the sides of the diagram in Figure 2.3 we can see that movement is continuous and that each 'move' consists of adding the number thrown on the dice to the number previously reached.

A simplified diagram which puts together the arrows, as the abstract version of the snakes and ladders, and the strip, instead of the square, might look something like the one in Figure 2.4.

This process of repeated 'abstraction' is the essential characteristic of good modelling.

The last stage in simplifying this model is to make the journey from the

Figure 2.3 The pattern of movement.

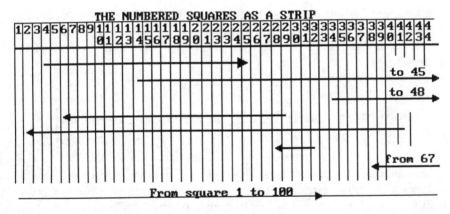

Figure 2.4 The simplified *visual* picture.

visual model to one which is purely numeric. Even the arrows and the strip are not absolutely essential since we can say, of any one 'move': if I reach 100 I have won; otherwise, at the simplest, the new P(osition), often indicated as P' will equal the old position, P plus the throw of the dice. This is fully represented by

$$P' = P + T(\text{hrow})$$

2.3.3 Only if P' is at the foot of a ladder or the head of a snake is there any complication, a 'slide' from that square to another. All such slides can be described by: if my counter lands on a *cause* number (or square) it ends up on an *effect* number (or square). We already have a fine mathematical tool for holding pairs of numbers like these. We call it an 'ordered pair' and the relationship between any such pair of numbers is preserved by our always dealing with them in the same sequence, in this case *cause* followed by *effect*.

2.4 The matrix model

2.4.1 When we have a number of such ordered pairs the best way of keeping them all in place is to put them into a *table* or *an array* or, as it has known to mathematicians for generations, a *matrix* (matrices in the plural). (There is no essential difference between any of these names and in computing we have simply fallen into the habit of making rather more use of 'table' and 'array'.)

2.4.2 Our final data model is a two-row matrix; a row of cause numbers and a row of effects. If the result of any throw is to be found in the top row the actual position achieved is to be found in the corresponding column of the bottom row.

In the case of our board:

$$\begin{bmatrix} 41 & 37 & 4 & 29 & 14 & 32 & 34 & 67 & 90 & 92 & 58 & 79 & 99 & 86 \\ 2 & 3 & 25 & 6 & 45 & 28 & 48 & 38 & 52 & 54 & 76 & 59 & 85 & 96 \end{bmatrix}$$

This collection of ordered pairs is just as I read them from the board. Usually a data set like this is handiest if it is ordered, i.e. sorted into a sequence, by using a 'key' of some sort. The key to events in the present case is the 'cause' number and we might better present the array as:

$$\begin{bmatrix} 4 & 14 & 29 & 32 & 34 & 37 & 41 & 58 & 67 & 79 & 86 & 90 & 92 & 99 \\ 25 & 45 & 6 & 28 & 48 & 3 & 2 & 76 & 38 & 59 & 96 & 52 & 54 & 85 \end{bmatrix}$$

and we can then examine it, after each throw, in sequence.

2.4.3 We can think of a matrix as a number of rows and columns of data-holders, or 'cells'. As is so often the case with mathematics we turn to the shorthand of symbols rather than using lengthy descriptions. For matrices we designate the matrix by a capital letter, let us call ours 'C', and each of the data elements stored in the matrix is identified by giving its position by its row and column address. We designate these by using a small '*c*' for the element and small numerals set at the foot of the '*c*' on the right-hand side, *subscripts*, to indicate, as an ordered pair, first the row number and then the column number of the *address* of the *cell*.

2.4.4 Because it is accepted practice to use '*i*' as a variable name to denote the row and '*j*' to denote the column of an address the general form of cell address is given by c_{ij}.

In the present case, the last ordered pair in our sorted matrix, 99 and 85 would be designated as $c_{1,14}$ and $c_{2,14}$, respectively and the fourth pair, for example, would be $c_{1,4}$ and $c_{2,4}$.

This notation may be familiar to you but if it is not you should make sure, by practising on our model, that you can use it freely because it is of real importance.

2.4.5 The final stage is to express our model as the core program segment, i.e. making one move for one player. We shall eventually need to allow for more than one player and if we set up a smaller array with '*n*' cells for '*n*' players then each player's position on the board, initialised to zero, will be designated by P_1, P_2 and so on.

Working out the model for one player only, using a pseudo-random number generator instead of a die, we have something like

```
DO WHILE more moves to be made
j ← 1, T ← RND(6), P₁ ← P₁ + T
   IF P₁ = 100
      display 'You have WON', END
   ENDIF
       DO WHILE j < 15
          IF c₁ⱼ = P₁ THEN P₁ ← c₂,ⱼ
          ELSE j ← j + 1
          END IF
       END DO
   display P₁
END DO
```

Once we are sure that this, or something similar, actually works we are ready to begin working 'outwards' from the core segment to add all the other embellishments, including the beautifully coloured graphics or the exciting sounds, which will bring the game to life on the computer screen.

2.5 A numerical model

2.5.1. Often, in practice, we find that quite complicated situations may be reduced to comparatively simple numerical models.

2.5.2 A very early mathematical model was developed by the great Italian mathematician Leonardo Fibonacci some 700 years ago. He was conducting a scientific study of how some animals multiply. We will follow his modelling process.

Let us assume that a female rabbit comes to breeding maturity and has her first litter at two months of age. She is able to produce a further litter every month thereafter. If the worst case* assumption is that only one female from each litter survives to breed how many female rabbits will this generate in one year?

2.5.3 In the first instance we might illustrate this situation by a traditional family 'tree'. Tree models show changes, developments, or generations on each successive level, i.e. as trunk, bough, branch, branchlet, twig, leaf, with as many 'branchings' as the model needs.

If we illustrate our rabbits by drawings, which are themselves, of course, models or abstractions, we find the tree developing like that shown in Figure 2.5.

This, of course, would become very elaborate after a very few months and we clearly need something less time-consuming to draw and which takes up less space.

2.5.4 The remedy varies from one application to another but we usually represent each 'branch' by a pair of lines which 'grow apart' from each other at an angle.

2.5.5 For the present purpose I have chosen to use a single line (segment) to represent our 'standard unit' of one female rabbit.

2.6 A tree model

2.6.1 The result is a tree diagram (as illustrated in Figure 2.6) which has been converted, almost completely, into an abstract visual representation of our information. Such tree models have many uses and we shall examine them more thoroughly later in the book. In the present case any pattern which there might be is rather obscure but we *can* see that growth is at an increasing rate.

*Scientists often assume the 'worst case', i.e. what is the most pessimistic view of a situation.

Figure 2.5 The rabbit tree.

2.6.2 Fortunately this is another situation where we can go further and extract a numeric model which may clarify matters. We first look at the month-by-month record and *count* all the rabbits in our diagram. Whenever we reach this stage in the development of a numeric model we shall usually find it profitable to **tabulate** the data, i.e. put them into a table. The next step, in looking for any patterns which may help us, is to look at the differences between successive pairs of numbers, in this case the size of each month's increase in the number of rabbits.

Figure 2.6 The abstract tree.

2.7 A table model

2.7.1 Setting these results into a table we get Table 2.1.

Table 2.1

Month	Start	1	2	3	4	5	6	7	8	9
Rabbits	0	1	1	2	3	5	8	13	21	34
Increase			0	1	1	2	3	5	8	13

2.7.2 Can you complete the table for the 12 months?

2.8 A model as a sequence

2.8.1 This pattern of growth in the numbers of rabbits is, when we study it, completely predictable and any such set of numbers is known as a **sequence.** This is another of those words which has a special meaning in mathematics. It means not only that the numbers, or 'terms' as they are often called, follow one another but that we can always say just what the next number in the sequence will be. This particular model actually describes many thousands of natural growth patterns and is renowned as the 'Fibonacci sequence', after its great discoverer.

2.8.2 A program to model and develop this sequence is fairly straightforward:

$S_1 = 0, S_2 = 1, i = 3$
Accept T; [number of terms needed]
   Do until $i = T$
     $S_i \leftarrow S_{i-2} + S_{i-1}$
     write S_i
     $i \leftarrow i + 1$
   END DO
END

2.9 Conclusions

The process of modelling requires a knowledge of a range of mathematical techniques but is, nevertheless, one in which we can become competent by following a consistent routine. For the computer the most elegant model is not the first requirement. Developing one which is true, robust and which *works* is our first concern; the elegance can come later as we become more experienced and develop a pride in the *quality* of our models.

Exercises

(1) Write a mathematical model, *not* pseudo-code, of a process to calculate *P*, the selling price of any item, which includes the wholesale price plus a 30 per cent mark-up for overheads and profit plus an *overall* government tax of 3 per cent.

(2) $V = \frac{4}{3}(\pi r^3)$ – write a pseudo-code model to find *V* where π is a constant, *r* is accepted from the keyboard and finally *r* and *V* are printed.

(3) Develop a pseudo-code model of a stock valuation procedure which will include, for each item in the stock-list, RECEIPTS, ISSUES and STOCK-HOLDING, day-by-day, with both the quantities *and* prices of receipts and issues, varying from day to day.

(4) Expand the model of the snakes and ladders game to allow for up to six, named players.

(5) The sequence in 2.8 produces an interesting variation when it generates the set *F* in which $f_n = f_{n-2}/f_{n-1}$. Develop and dry-run a pseudo-code model of this.

(6) Develop a model to calculate lorry transport costs per kilometre if, for each vehicle:
   (a) the distances run each month are highly variable
   (b) fixed costs are $700 per month
   (c) variable costs are $2 per km
   (d) depreciation is $500 per month.

(7) A new payroll tax has been introduced. It is to be collected by the employer and paid by him to the government. Employees and employer contribute jointly with the former paying 1 per cent of gross earnings per week or per month depending on how frequently they are paid. The employer must add $1\frac{1}{2}$ times the amount contributed by the employee.
   (a) State this as a mathematical model.
   (b) Write a pseudo-code routine to be added to the current payroll program in due course.
   (c) Devise appropriate test data and dry-run your routine. Debug if necessary.

(8) Model, in pseudo-code, a range-check routine to check the total of every tenth employee's pay to ensure that it is not a negative amount and also that it does not exceed $1000. Each full payroll run of this routine should start with a different member of the set {the first ten employees on the payroll} chosen at random.

(9) (a) Develop a tested pseudo-code model which will calculate and print a *two-way* timetable for local trains starting at 'A' and 'E', respectively based on the following data. The trains are to depart at 20-minute intervals from 7 o'clock in the morning to 7 o'clock in the evening. They run at an average speed of 40 km/hour and the

distances covered are: A to B, 10 km; B to C, 8 km; C to D, 14 km; D to E, 16 km. The timings must allow for station stops of 2 minutes at B and C and at least 8 minutes at A and E.

(b) What is the minimum number of trains needed to run the service if one is always to be kept in reserve?

(10) An electricity supplier offers customers a choice of three tariffs:

(a) $0.28 per unit no matter how much electricity is used.

(b) A fixed charge of $60 for up to 250 units and $0.35 per unit for the rest.

(c) $0.30 per unit for the first 400 units and $0.25 for any units above 400.

Write the necessary pseudo-code segments to find the total charge, C, for U units consumed, depending upon the tariff applied in any particular case.

3

Sets and their applications

Objectives

By the end of this chapter you should be able to:

❑ understand the logical requirements of a set
❑ define, enumerate and specify sets
❑ solve simple problems using sets
❑ understand the relations between sets
❑ derive simple mappings and functions.

A reminder: in the Introduction we saw that to be a 'proper' mathematical set a collection must be *defined* and each of its elements must be *unique*. (If, in practice we have a collection which *is defined* but includes items which *are* duplicated we sometimes refer to it as a 'bag' and it cannot be dealt with by the rules of sets.)

3.1 Defining a set

3.1.1 When discussing a payroll model in the previous chapter we looked briefly at the test data which would be needed to validate H, the hours worked. We saw that both H and the rate of pay could, for the sake of improving processing accuracy, be expressed as integers. In the case of H we did not make a suitable conversion at the time; let us now do so and consider the implications for valid inputs to H and for the essential test data.

The hours were to be recorded *only in completed quarters* so that someone who worked, say, for 7 hours and 18 minutes would be credited with $7\frac{1}{4}$ hours only. In order to use integers we should convert all values for H to the equivalent number of *quarters* and, at the very end of our calculations, divide the calculated pay by four to arrive at a final total. Our $7\frac{1}{4}$ hours equals 29 quarters.

3.1.2 For practical programming purposes we must now examine the implications of this. In one week there are 7 * 24 hours, 168 hours in all or 672 *quarters*. No-one can possibly work more than this. Similarly no-one can work a *minus quantity* of hours! (Although this sounds absurd it is entirely likely that someone, by simple human mistake, would sooner or later enter a minus quantity and we must therefore expect it and guard against it.) Our data range is, therefore, from 0 to 672, inclusive. In mathematical symbols we should write: $0 \leq H \leq 672$.

Since such expressions may not be immediately familiar to all of us let us translate it back into words: '0 is less than or equal to H, the number of quarter hours worked, which is less than or equal to 672'. This is still clumsy and may be better understood as, 'H must be at least 0 and at most 672'(*).

3.1.3 One of the most difficult things, for any programmer (or anyone reading a program listing), is to ensure that any conditions which control a program are logically correct. This is precisely where we can best use the language and principles of sets to help us clarify our thinking. Note that the proper time to turn to this branch of mathematics is *before* we start writing code, i.e. in the *design of our program,* as we seek to develop the core model into a practical system.

In the present case, by defining our data range as $0 \leq H \leq 672$ we are, without stating it *explicitly*, also declaring that *any other value is wholly unacceptable*. This is an aspect of logical expression which is not common in everyday discourse (discussion or exchange of ideas), and it is quite difficult to learn to think in such a rigorous way.

Unfortunately the computer is quite unable to tell right from wrong, logical from illogical, and we *must* learn to think like this if our programs are not to be riddled with errors of logic, as is all too common. Put quite simply; every time that we define a logical **is** we also, *automatically*, define a corresponding **is not**. What, then, *is not* within the acceptable data range for H? We can state the conditions for both H and **not** H in many ways. Making a list of what is acceptable is one way but any such list may quickly become too elaborate and too long for convenience.

3.1.4 The simplest way is to define H as a **set** and we can do this in several ways:

(a) By a verbal description such as that shown above by the (*). It is always possible to describe such situations fully in words if one is not comfortable with mathematical symbols but where logic is concerned, if we are to avoid any possible ambiguity or misunderstanding, this, can be a very long-winded process.

(b) By listing all the elements (members of the set); this is fine for small sets but becomes both cumbersome and prone to mistakes for sets of any but the smallest size. See Figure 3.1.

(c) By abstract mathematical symbols; this is, without doubt, the most compact description of a set. We will look carefully at this symbolism presently and investigate all the symbols which we use in order to learn this most valuable shorthand as we go along.

(d) By Venn diagram; this, for most people is clear and easy to work with yet provides us with a very powerful tool, as we shall see. Its limitation is that it is often difficult to display large numbers of elements except as quantities.

3.1.5 What is our 'universe of discourse' (usually given the mathematical symbol $\mathscr{E}$), in this case? The only valid data for our model is the *number* (of quarter hours) assigned to H. The variable is *numeric* and we should usually see a run-time error message if we tried to input anything but numeric data. (Nevertheless we cannot take this for granted and must check the particular compiler or interpreter which we shall be using to ensure that it does.)

Figure 3.1 The Venn diagram can get crowded.

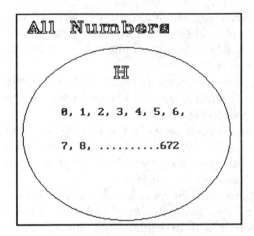

Figure 3.2   Our small data set in a very large 'universe'.

3.1.6 In this case our $\mathscr{E}$, the universal set, is the *whole world* of numbers; of which we are prepared to accept as valid only *0* or *positive whole numbers up to and including 672*. See Figure 3.2.

What must be rejected is all which is *outside* this small set of valid data.

3.1.7 A Venn diagram will help us see the whole picture. What is *invalid* is *all* other types of number including integers <0 and all real numbers which are *not* positive integers ≤672. Since our test data must always include examples which should *not* be accepted, as well as those which will show that the program does run successfully, we must select sample values from $\mathscr{E}$ also.

3.1.8 The entire area of the universal set outside a named set is known as the **complement** of that set and is designated by a prime (′) symbol. Our $\mathscr{E}$ consists of *H* together with *H′*.

3.2 Relations between sets

3.2.1 The small part of $\mathscr{E}$ which *is* acceptable to us is, however, part of several other sets of numbers and the less confident we are with traditional mathematics the more valuable we are likely to find it useful to detail *H′* more carefully as in Figure 3.3. Here we begin to clarify the complex relationships within $\mathscr{E}$ and identify our valid and *invalid* data more readily. Remember that everything *outside* any set is 'NOT', in our case *not* valid.

3.2.2 Whenever we analyse a problem involving several sets it is best to start from the innermost set and work outwards.

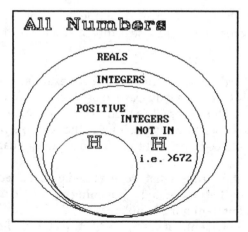

Figure 3.3 Sets within sets.

Note: a set is indicated by placing the name, or the list of is elements (it is not essential that they be ordered), between 'braces' {1, 2, 3, 4}.

3.2.3 Sets which are wholly contained within others are known as **subsets** and designated by the special symbol, $\subset$.

3.2.4 Our current diagram tells us that: $H \subset P$ ({Positive Integers}. But all of the latter which are >672 are *invalid* for our task.) $P \subset I$ ({Integers}. But only a subset of positive integers is valid for us.) $I \subset R$ ({Real Numbers}. Any of which having fractional parts are invalid for us.) $R \subset \mathscr{C}$ (All Numbers).

3.2.5 We now have a logical model which indicates what data-testing we need to do in our validation procedures for H.

(Note that the new symbol $\neq$ signifies 'does not equal' or 'is not'. The programming language reserved word, INT(n), which we will also use in our pseudo-code, signifies 'take the whole number part only of 'n'.)

```
IF H ≠ Numeric then INVALID
   ELSE IF H − INT(H) > 0 then INVALID
      ELSE IF H < 0 then INVALID
         ELSE IF H > 672 then INVALID
   ELSE H is VALID
   ENDIF
```

Make sure that you can see what each line is meant to do and, as always, check very carefully to make sure that the segment works and that it accurately represents the logical structure of sets which we have been developing.

3.2.6 This is a good point at which to use our example to introduce some other new symbols so that we may learn to use them to avoid having to write verbal descriptions. That, after all, is their main value but equally important is the fact that the mathematical symbols mean the same the whole world over whilst words have to be translated from language to language!

3.2.7. To clarify still further what we discussed earlier a set, no matter how large or small, may be designated by **naming** it, for example {the set of integers}, or by a **letter**, in our example, *H*.

We may **list** the elements of a set, say *A* = {cat, dog, horse, wolf, tiger, panther, lion, elephant}, if it contains only a few **elements**. We could then say, for example, 'tiger ∈ *A*'. (Tiger is an *element of* the set which we have labelled '*A*'.)

Note very carefully that we are not implying that the set '*A*' designates any particular **attribute** of its elements; the set is *not*, for example, *necessarily* a set of mammals although it would be, without a doubt, correct to say, if the images in our illustration of this set, in Figure 3.1, represented *real* animals, as opposed to, say, furry toys, that '*A* ⊂ {mammals}'.

You should check and discuss all such statements to make sure that you and your fellow students follow the *logic* of the situation and are learning not to jump to conclusions.

3.2.8 Any small set, such as *A*, can also easily be **enumerated** or **numbered**, i.e. counted. We often use '*n*' to stand for any number, known or unknown, in mathematics and we can say that *n*(*A*) = 8 or that *A* has eight elements. Very large sets may be impossible to number even if they are **finite**, i.e. have a fixed number of elements whilst many sets are **infinite**. Discuss what this means and decide what effect the *size* of a set, of a file perhaps, may have in practical computing terms.

3.2.9 Let us, for its designation in **abstract symbols**, return to our example. Our set, *H*, can be designated as:

$$\{0 \text{ together with the positive integers} \leq 672\}.$$

This is still rather clumsy and we are better writing {0, 1, 2, 3, ..., 672}(*see below).

The more purely symbolic form you may find written in mathematics textbooks as:

$H = \{0;h_i:1 \leq h_i \leq 672, h_i \text{ integer}\}$ – or something similar.

Do not be intimidated by such statements! We can retranslate them back into plain English a step at a time:

H designates a set which consists of zero and individual elements, h_i, which are at least 1 and not more than 672 and all of which are integers, or whole numbers.

There are many occasions when it is useful to abbreviate our descriptions in this way but if in doubt, use the 'listing' method, shown by *, earlier.

3.2.10 There are a number of other symbols which we need to use and this is perhaps a good time to list them:

=	is equivalent to
≈	is approximately the same as
≡	is the same in all respects as
$\mathscr{E}$	the universal set, or universe of discourse.
{...}	the set ...
∈	the item to the left is an element of the set to the right (of the symbol)
∉	is *not* an element of
⊂	is a true subset of
⊃	has, as one of its subsets
∪	the union of sets (e.g. $A \cup B$) puts together all the elements of A with all those of B
∩	the intersection between (e.g. $A \cap B$) groups all elements which are in *both A and B*
/	this sign makes any other symbol into 'NOT'
∞	infinity – 'goes on for ever'.

3.3 More complex relations

3.3.1 There are many ways in which we can represent sets by Venn diagrams and we should pause to look at a few typical examples. Some of the relationships are indicated in Figure 3.4 on the next page. See if you can fully explain the logical story told by examples (a)–(h), in words and, if possible, in symbols.

3.3.2 You should, with your colleagues, select various 'universes of discourse' and explore the sets within them. For example, the universe of four-sided, straight-line, mathematical shapes which includes rectangles, squares, oblongs, parallelograms, trapezia, rhombuses, irregular shapes, and perhaps others! Can you see why we need the word 'oblong', which many people wrongly think of as slang?

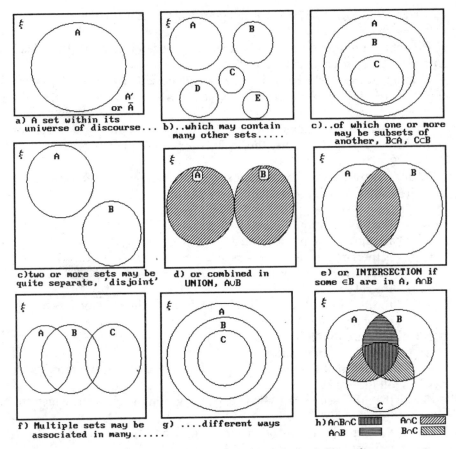

a) A set within its universe of discourse...

b)..which may contain many other sets.....

c)..of which one or more may be subsets of another, BⲥA, CⲥB

c)two or more sets may be quite separate, 'disjoint'

d) or combined in UNION, AⱶB

e) or INTERSECTION if some ∈B are in A, AⱶB

f) Multiple sets may be associated in many......

g) ....different ways

h) AⱶBⱶC ▨ **AⱶC** ▨
AⱶB ▬ **BⱶC** ▨

Figure 3.4   There are many possible variations on the basic Venn diagram.

3.4 Practical set models

3.4.1 An extremely important application of sets, in computing, is that governing an organization's **database**, that is to say the total storage system for all the information needed to run the organization.

This vast universe of discourse will contain three major subsets: the first, the information needed for the day-to-day routine activities of the organization; the second, that which is needed for middle management, tactical, decision making and finally, that needed for top-level, strategic planning and decision making.

3.4.2 It is imperative that the structure of the database be rigorously logical and the systems analyst must be able to establish all the subsets and the relations within and between them.

3.4.3 As an example of how involved such relations may be a stock or parts list is an especially complex example of multiple subsets, as in Figure 3.4(g).

3.4.4 As a practical task try to create a stock list, with all its appropriate subsets, of the stationery and equipment which is needed by your present study class just for the present lesson.

3.4.5 What follows from this taxonomy of stock is that individual items must also be named, consistently, according to strictly logical description rules so that everyone in the organization knows into which subset any particular item must be placed. Once again, try a number of examples, such as 'nail, flat-head, wire, 75 mm long'. What does this tell us? That, within a set of fastening devices (which might include screws, bolts, rivets, etc.) there is a subset, nails, of different head types, made of different materials and of different lengths. There may well be flat-headed nails made of other material; there will be a variety of nails of length 75 mm, so there will be a number of intersections. What we need at this stage is to explore such situations with diagrams and *careful thinking* rather than any elaborate calculations.

3.5 Sets and domains

3.5.1 When we look at a Venn diagram we see that the boundary lines of the different sets create *regions*, which we generally refer to as **domains**, within the universal set.

In Figure 3.5, for example, there are eight domains in all; let us identify them clearly and then label them in formal symbols. (For the time being we will continue to use the '−' symbol for 'less' or 'excluding' but we *will* use the '∪'

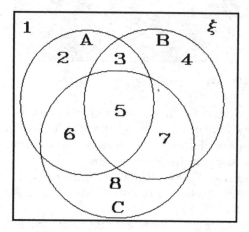

Figure 3.5 The domains identified.

symbol for combining sets. Note that it is always easier to read such sequences of symbols if we keep the letters in alphabetical order.)

The relevant descriptions are:

(1) $(A \cup B \cup C)'$
(2) $A - ((A \cap B) \cup (A \cap C))$ or, in words, 'A less what is in A but *is also* in B and C'.
(3) $A \cap B$
(4) $B - ((A \cap B) \cup (B \cap C))$
(5) $A \cap B \cap C$
(6) $A \cap C$
(7) $B \cap C$
(8) $C - ((A \cap C) \cup (B \cap C))$.

3.5.2 We shall need, for more precise labelling, two other symbolic representations. When we talk of the set A we describe what is **in** A. But we also imply that the elements in any other region or regions, however complex, are **not in** A, as shown in Figure 3.6.

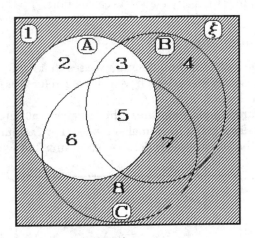

Figure 3.6 The shaded region is 'NOT *A*'.

3.5.3 This, in the case of a single set within its universe, we have already seen to be the *complement of A* or *A'* or whatever is needed to *complete the whole* (universe of discourse). 'However, we often need to identify, as NOT(IN)*A*, one or more of the elements in other domains'. This may be (much) less than the complement of A and we use the representation or, for example, means 'what is **not** in the intersection of A with B'. See Figure 3.7 opposite.

3.5.4 It will also occur quite frequently that a set will be empty, i.e. it will have *no* elements.

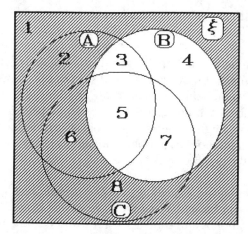

Figure 3.7 NOT *B* is shaded.

We could, for example, quite logically define {human beings over 3 m tall} and find, within any particular population that the tallest person is 2 m in height.

Similarly there will be sets within a database which, from time to time, are empty. Most shopkeepers will hope that the set of 'bad debtors' will *normally* be empty but the set must still exist within the database.

3.5.5 We call such sets 'null' with the symbol Ø. In rigorous logic, since any collection of nothing is exactly the same as any other collection of nothing we speak of *the* null set rather than 'a' null set.

3.6 Solving problems

3.6.1 Before going any further let us look at the kind of problem which can be solved using nothing but Venn diagrams.

One hundred people were asked which of three radio stations they had listened to at any time during the previous evening. They said that they had listened as follows:

    65 to station A

    45 to station B

    55 to station C (can you suggest how it can be that these numbers total far more than 100?).

Several people had listened to more than one station at some time during the evening:

    22 to both A and B

    32 to both B and C

    35 to A and C and 12 people had listened to all three.

What story do the 'listening' figures actually tell us and were there any who didn't listen to any of the three stations?

3.6.2 When we see a situation like this, including lots of numbers, most of us are tempted to pick up our pencils and start doing 'sums'. As always, let us look and think a bit before we start *doing* anything.

Since the numbers we are given actually add up to far more than 100 ordinary arithmetic might be a bit confusing! A further look shows that we are dealing with a total of 100 people who fall into three 'categories' or into combinations of those three categories.

3.6.3 Whenever we are 'classifying' data we should first look to see if those groupings are or can be sets. What we have in this case is the set of people who listened to station A and similar sets for B and C. We can now model this problem, or any others like it, by a Venn diagram.

LET A = {people who listened to station A}

LET B = {people who listened to station B}

LET C = {people who listened to station C}

and since there are some people in each of the possible combinations of A, B and C we need a 'three-set intersection' diagram into which we can put our data. Figure 3.8 shows us what it looks like.

The letters denote the sets and the numbers in heavy type the domains, just as in our earlier example, Figure 3.5.

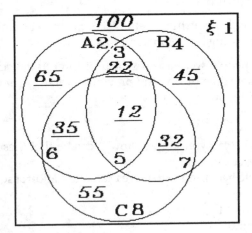

Figure 3.8   The 'raw' numbers allocated.

3.6.4 Our actual data are inserted as they were originally listed but we have not yet adjusted them to total 100.

3.6.5 Here is where we start from the centre and work outwards. The people in region 5 are also in regions 3, 6 and 7. Can you see why?

Those people who listened to *A, B and C* are *also* included in the numbers of those who listened to either A and B; A and C or B and C. Similarly the number of people reported as listening to A includes those who listened *also* to B, *or* to C, *or* to *both* B *and* C.

The rules of sets, however, dictate that there may not be duplicates of any element in a set so those in domain 5 must be subtracted from each of domains 3, 6 and 7.

In exactly the same way the 10 + 12 + 23 people already represented in domains 3, 5 and 7, respectively are all in *A* and their total must be subtracted from the total number of elements in *A*, or *n(A)*.

So 65 – (10 + 12 + 23) leaves 20 people *who are in A only* Finally we see that 'C only' is null and that 12 people did not listen to any of the three stations. This is shown in Figure 3.9.

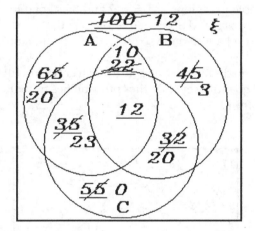

Figure 3.9 'Cancelling' the duplicates.

3.6.6 This is our objective in the model, to convert the rather muddled information given in the narrative into true or proper sets. Once we have done this there is nothing which we do not know about the *numbers* concerned. Perhaps we should, however, remind ourselves that we know *only* about the numbers. The limited information which we have tells us nothing about the *quality* of any of the radio programmes, for instance.

3.7 The domains in a probability space

3.7.1 This model of a problem, by allocating *all* the information to separate domains and refining it, actually gives us another very useful tool.

3.7.2 We have created, almost as a by-product, a 'map' of probabilities or a 'probability space', bounded or limited by ℰ. We shall be exploring the ideas of probability much more fully in a later chapter, but let us just glance at the story our 'map' tells us. For example:

If we met one of the 100 people represented by our ℰ what is the likelihood that he or she would have listened to:
   (a) station A
   (b) both A and C
   (c) none of the three stations
   (d) at least two stations?

3.7.3 (a) Twenty of the 100 people listened to A only so the likelihood of our one person having done so is 'twenty out of one hundred', 20/100 or 'one in five'.
   (b) All the people in regions 2, 3, 5, 6, 7 and 8 listened to either *A or C* but only those in regions 5 and 6 listened to *both*. So there are 35/100 chances of our selected person having done so (or 7/20). Look at Figure 3.10 below.

3.7.4 You will, no doubt, now be able to answer **(c)** and **(d)** for yourself. Try also to label and quantify several other probabilities from our map; there are many to choose from.

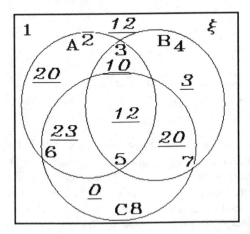

Figure 3.10   A probability 'map'.

3.7.5 Such probability models are rather easier to work with than many others and are well worth practising.

3.8 Other logical relations

3.8.1 Before we leave our Venn diagram models we should look at one other aspect of notation which we shall be using extensively in a later chapter.

3.8.2 For purely symbolic logic descriptions and operations, as well as in programming logical decision-making statements, we use the words AND, OR, XOR and NAND to describe relations between sets. Figure 3.11 shows what they convey.

(a) the shaded area shows the domain which contains all those elements which are in *both* A *and* B
(b) the shading identifies everything which is *either* A *or* B
(c) here we have what is in A *or* B *but not in both*
(d) indicates what is *not in* A *and not in* B.

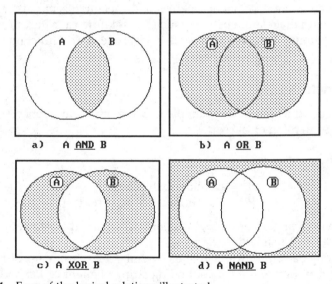

Figure 3.11   Four of the logical relations illustrated.

3.8.3 As always we should take a real example and practise using this notation. We will often find it valuable to test programming statements such as:

DOWHILE $A < B$ or $A < C$ and $A > D$

by modelling them as a Venn diagram to make quite sure that our statement actually does what we want it to do. Look at this one carefully; does it, for instance, need any brackets to clarify its purpose?

3.9 Mapping

3.9.1 There is an extensive range of relations between sets which are between the individual *elements* of two or more sets rather than between their wholes. These are best explored using ordered listings rather than diagrams.

3.9.2 For example there is a clear relationship between: {cat, dog, horse} and {foal, kitten, pup} but we may see the relationship better if we *order* our data and set out the model as:

$$P = \{\text{cat, dog, horse}\}$$
$$O = \{\text{kitten, pup, foal}\}$$

when each element in the 'top' set may be linked visually to its corresponding element in the 'lower set' and we can clearly see that the relationship is that of 'parent to offspring'.

3.9.3 Such a correspondence between the elements of sets is called a 'mapping' and the relation between such sets is known as a 'function'. Within all programming languages there is a clear need for functions and the procedures for dealing with them.

3.10 Functions

3.10.1 The function, in the example in 3.9, might be described as 'parenthood', or something of the sort.

3.10.2 In another example:

$$S = \{\text{cat, home, ship, banana}\}$$
$$P = \{\text{cats, homes, ships, bananas}\}$$

which is a mapping of the 'string' variables which are the elements of the set S on to set P by the function '$P_i' = 'S_i' + 's'$.

Translating back into plain language we might say, 'from any string in the set, S, we may derive the corresponding string element of P by adding the string "s"'. Within all operations on strings the process of appending, or joining, one string to another is commonly called *concatenating*. In computing we often have to deal with text and string functions so that such concatenations of strings are routine operations.

3.10.3 For the moment it may be a little easier to look at some number sets and their mappings and functions.

3.10.4 Consider the following

$$\{0, 1, 2, 3, 4, \ 5, \ ... \ \infty\}$$
$$\downarrow, \downarrow, \ \downarrow, \downarrow, \downarrow, \ \downarrow,$$
$$\{0, 2, 4, 6, 8, 10, \ ... \ \infty\}$$

This mapping shows, in very formal symbolism, a situation with which we have all been familiar since we were quite young. What is the function?

3.10.5 This shows a **one-to-one correspondence,** in other words each element of the 'parent' set relates to a corresponding element of the 'derived' set. Here is another kind of relation:

$$\{1, \ 1, \ 2, \ 3, \ \ 5, \ \ 8, \ \ 13, ... \ \infty\}$$

$$\{ \ 1, \ 0.5, \ 0.66, \ 0.6, \ 0.625, \ 0.615, \ ... \ \infty \ \}$$

3.10.6 This is a **two-to-one correspondence** because it takes *two* elements of the parent set to generate *one* corresponding element of the derived set.

3.10.7 Once again you should identify the mapping function. Do you recognise this parent set, which we met in Chapter 2?

3.10.8 We may meet other variations such as 'one-to-two' or 'one-to-many'; there is no practical limit.

3.11 Numeric functions

3.11.1 In different programming languages we shall find different ways of defining or declaring functions. In our last example, for instance, we might find it coded as something like: DEF Fn $d(p) = (p_i/p_{i+1})$.

3.11.2 If we, as always, translate this back into plain language it says something like, 'define the function to generate any individual element, d_i, of the derived set, from the corresponding element of the parent set we divide p_i by the next element of the sequence, p_{i+1}'.

3.11.3 For example, to find the fifth element of the derived set we take p_5, which is eight, and we divide it by p_6, which is 13. The fifth derived element is, therefore 5/8 or 0.625.

3.11.4 A very common way of writing such functions in conventional mathematics is as, for example $y = 2x + 7$.

3.11.5 In such traditional mathematics exercises we are often required to take a set of possible values for x and find the corresponding values for y.

3.11.6 If we set it out in the programming code form which we used in our last example we might restate it as:

$$\text{Def fn } y(x) = 2x + 7$$

which reads, 'to find any y_i we take the corresponding x_i, multiply it by two and then add seven.'

3.11.7 When we are working such examples in the traditional way we may set the values into a table instead of in the form

$$X = \{\infty, \dots, -4, 0, +4, \dots, \infty\}$$

so that we write something like:

IF		$x =$	-4	0	$+4$
THEN	$y = (2^*x) + 7 =$		-1	$+7$	$+15$

3.11.8 What we have, although we may never have seen it in that light before, is a set of x values mapped, in a one-to-one correspondence, onto a set *of y* values by the function $2x + 7$.

3.11.9 We may, in a textbook, be told that the x values are limited to integers; or are such that $-10 \le x \le +10$. This does not necessarily make it easier to select possible values for x, especially if we are testing a program routine.

3.11.10 In real life it may be the systems analyst or ourselves, as programmers, who must *set* these boundaries. We must *know* what is **in** the permitted set of values for x and what is **not**. Unless we are given a specific range of values for x the set of possible values may well be infinitely large and any 'test' values which we may choose, in order to examine or display the function, are only a small subset. We must always remember that our subset is just a *representative sample of the possible values* and that the choice of test data, in particular, must always reflect that reality.

3.12 Conclusions

In a rather similar way we have seen only a subset of the applications of sets in this chapter.
We shall see others in practically every subsequent chapter.

$$n\{\text{applications of set theory}\} = \infty\ !$$

Exercises

(1) What are the two conditions governing a proper set?

(2) List three collections of entities which constitute proper sets and three collections which are not. Give your reasons in each case.

(3) Identify from the following groups those which are not proper sets and give your reasons:

(a) people aged over 16 years
(b) fat people
(c) debtors owing more that $10,000
(d) customers living in Hong Kong
(e) employees who work hard
(f) Manchester United fans
(g) a company's stock list
(h) women who prefer brand 'X' detergent
(i) the test data for validating a date
(j) real numbers.

(4) Draw Venn diagrams to illustrate:

(a) $\{1,2,3,4,5\} \cup \{4,5,6,7,8\}$
(b) $\{1,2,3,4,5\} \cap \{4,5,6,7,8\}$
(c) $\{1,2,3,4,5\} \cap \{4,5,6,7\} \cap \{5,6,7,8\}$.

(5) A poll of 300 people boarding a Singapore flight at London airport showed that many were continuing their journeys to other destinations. One-hundred-and-sixty were visiting Singapore; 160 were going to Malaysia; 110 were going to Hong Kong. Fifty were going to Malaysia and then on to Hong Kong and 30 of these were first spending time in Singapore. Fifty were merely changing flights in Singapore and going only to Hong Kong and 80 were going only to Singapore. How many were merely changing flights at Singapore to visit only Malaysia?

(6) Of 25 programmers eight knew *only* FORTRAN, five knew *only* C++ and 1 knew *only* Pascal. Of the rest six could use both C++ and Pascal, nine knew both C++ and FORTRAN whilst six knew FORTRAN and Pascal. Find how many could work in all three languages. A Venn diagram will undoubtedly help you but your answer should be a number only.

(7) Find the set of values which results from mapping the function n^3 from $\{n: integer\ 0 < n \leq 10\}$.

(8) What are the mapping functions of:

(a) $\{0,1,2,3,4\} \rightarrow \{0,1,8,27,64\}$
(b) $\{33,34,35,36,37\} \rightarrow \{1,2,3,4,5\}$
(c) $\{J,F,M,A,M,J,J\} \rightarrow \{31,28,31,30,31,30,31\}$
(d) $\{0,1,2,3,4,5\} \rightarrow \{2,4,6,8,10,12\}$
(e) $\{3,5,9,13\} \rightarrow \{17,15,11,7\}$

(9) Describe the sets defined by the relations:
    (a) {people over 1.5 m tall} $\cap$ {males}
    (b) $A \cup B \cup C$
    (c) $\{A \cup B \cup C\}'$
    (d) A XOR B
    (e) A NOR B
    (f) $\{x\colon -10 \leq x \leq 10\}$

(10) List three reasons why a computer database has to be organized as a universe of discourse with sets and subsets.

4

Number sets and systems

Objective

After working through this chapter you should be able to:

❏ identify the different types of number sets and their applications.

4.1 Some number sets

4.1.1 We have already touched briefly on the existence of many different number sets, let us now explore this topic in a little more detail. As I work I see that there are six pieces of computer equipment on my table and I may express this in set notation as:

$$C = \{\text{computer equipment on my work table}\}, \text{ and}$$
$$n(C) = 6 \text{ or, the number of elements in } C \text{ is } 6$$

4.1.2 To what number set does this six belong? Any collection of *actual entities* has a **natural** size or quantity, whether or not we ever put it into words or symbols.

4.1.3 This has often been described, if somewhat clumsily, as the 'n-ness' of the group of entities. In strict mathematical language it may be described as the **quantum** of the group, *which is independent of the symbols or words which we use to describe it!*

With another example (shown in Figure 4.1), of horses in a field, the group has this *inherent* size, or 'number' which we happen, in English, to call 'five' but the word is only one way of describing that number.

Figure 4.1 The '*n*-ness' of horses.

4.1.4 The important point is that man has invented the *word*, but the *quantum* is a *natural* attribute of the group. This attribute remains the same whatever language the word itself may be in. Not surprisingly we describe such numbers as {natural numbers}.

What is also very important to us is that *all other numbers are man-made inventions*. For example we cannot, in nature, have half a horse. In other words the central idea, the *concept*, of fractions is man-made and entirely *un*-natural.

Equally the 'five-ness' of our group of horses does not, in nature, have the *label* 'five' much less is it identified by the numeral 5. The words, the numerals or figures and the column system are all simply parts of a system of notation by which we *describe* and *manipulate* numbers; a set of human inventions which have evolved into the commonly used system of today over many centuries.

4.1.5 From the very outset we must always distinguish between the **number**, the actual *quantum* or quantity, and both the **symbols** and the **number system** which we use to *represent* that number.

4.1.6 Why do we need the words and the symbols? First and foremost in order to communicate, store and manipulate numbers representing quantities or sizes of things.

4.1.7 A secondary, but quite different application is where we also use them to identify individual positions within a sequence.

For this chapter we are concerned firstly with the *types* of number which we use and secondly with the systems which allow us to express and manipulate those numbers. Computer systems allow us to economize on time and space by using the type of number best suited to any particular application. They also store and manipulate numbers according to several systems which are quite different from the one which we use in everyday life. In order to make best use of this power *we* need a much clearer understanding of the nature and qualities of numbers than do ordinary users of mathematics.

Once again it is the *understanding* of principles which is important rather than traditional skills at 'doing sums'.

4.1.8 For the purpose of labelling the positions of items in a sequence, *or as a unique identifier in a set such as a file*, for instance as a firm's customer numbers, we use {ordinal numbers}. In words, these are the ones we *speak* as 'first, second, third, ...' even though we *write* them in numerals as '1, 2, 3, ...'.

4.1.9 All other, arithmĕtic, numbers are within {cardinal numbers}. In this case the word 'cardinal' means 'of first importance', since the first importance of numbers is to describe *quantities*.

4.1.10 We must be very cautious indeed here. In ordinary circumstances we can say, quite simple, 'we cannot, and must not attempt to, do arithmetic with ordinal numbers'. We cannot, for example, say that house number 2 + house number 3 = house number 5! We cannot even say that, because 5 − 2 = 3, there must be three houses between 2 and 5; clearly there would, at most, be two but there may be none! Such examples quickly reveal the absurdity of this false arithmetic.

4.1.11 In computing, however, there is no such simple separation of the ordinal numbers from the others. Within this special field such numbers may well *be* subjected to a range of arithmetic manipulations, for example, in the generation of check digits. We might also make equal subdivisions of a file by deriving true quantities, or *natural numbers*, from differences between customer serial numbers or dates, say.

4.1.12 The second problem for us is that within the ALU, as in storage, ordinals appear to be no different from natural numbers. They are, in fact, handled exactly as if they *were* ordinary positive integers. Once again it is we who have to understand the critical differences and ensure that, *at no time in our programming* can the ordinals be wrongly included in calculations. The computer doesn't know the difference; therefore *we must*!

4.2 Subsets of cardinal numbers

4.2.1 So far, as subsets of {cardinal numbers}, we have identified {natural numbers} and we have, in Chapter 3, touched upon {integers} and {reals}. How large is the set of natural numbers? What can it *not* contain? (Clearly, although it has the same elements, it cannot contain the {ordinal numbers}!)

These are the questions which we must always ask.

How many stars in the sky? Stars are certainly natural objects and have a natural '*n*-ness'. Astronomers, for their special study, do make estimates of

n{stars} – but if we started counting them we should have a quite hopeless task; there are simply too many and whatever number we reached in our counting someone could always say '... and one more'!

4.2.2 For ordinary mortals that is probably as good a working model of infinity as any. Theoretically we *could* count the members of any natural set but in many cases, such as the stars, we should always be able to say, 'and one more'. {Natural numbers} must then be infinitely large and if we were listing it we could write

$$\{1, 2, 3, 4, 5, ..., \infty\}$$

where the symbol before the right brace means infinity.

4.2.3 This set coincides with {positive integers} since all its elements are whole numbers only, without fractions, and they are all >0.

4.2.4 Zero is not, in any sense, a *number* at all; it is really the *absence of* number. It is useful, nonetheless to treat it as the origin of all counting and of the number line.

4.2.5 Nature has no 'minus' quantities; we cannot naturally have fewer than *no horses*. It is man who has invented the ideas of 'below sea-level; below zero; having an overdraft; owing a debt to someone; a negative balance on an account' and all the many conditions which we describe by negative, (positive), numbers.

4.2.6 So the set of *all* integers, as opposed to {positive integers} has {natural numbers} as one of its subsets. Thus:

$$\{natural\ numbers\} \equiv \{positive\ integers\}$$

but

$$\{natural\ numbers\} \subset \{integers\}$$

The two number lines are also quite different, as can be seen in Figure 4.2 and reflect the great extension of {integers} into the *unnatural*, negative integers.

4.3 Fractional numbers

4.3.1 If we need to deal with the *real* world, however, we may well need to measure, compare or account for *parts of units*. We could not, for example, measure distances only in whole metres – think of what our problems would be if we could only buy shoes which were 1 m long!

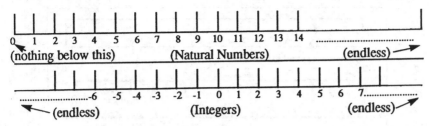

Figure 4.2 The two sets as number lines.

4.3.2 Unless we led very leisurely lives it would also be extremely inconvenient if all we knew of the departure time of a train was that it is, 'sometime between three and four o'clock in the afternoon'; or if, even for a family of two people, we could only buy tea in complete kilos.

4.3.3 In the real world we simply have to subdivide units and we express all *measurements* in quantities which reflect this fact. This means that, although we can *count* anything in natural numbers and express shortages of *items* in negative integers we must have a third set of numbers in which we can express, *not only* whole quantities *but also* any parts of units, no matter how small.

4.4 Real numbers

4.4.1 This need reflects the demands of the *real* world and we call the resulting numbers {REAL numbers}.

4.4.2 Let us take a magnifying glass to the number line as shown in Figure 4.3 and look at what, in the *real* number world, comes into the gap between *any pair* of integers.

Figure 4.3 'Closing in' on one part of the number line.

In Figure 4.4 overleaf, we see three 'stages' of magnification of the 'gap' between 0 and 1 on the **integer** scale which holds many, indeed an infinite number of, other values on the **real number** scale. In each case the thick line above the scale denotes the quantity which we are seeking to identify. At stage (a) we see the greatest degree of precision which we can arrive at using

Figure 4.4   Filling in, or opening out, the intervals between integers.

integers. Stage (b) shows us that we can considerably increase the precision if we subdivide our *unit* into ten equal parts or fractions, in this case, tenths.

By subdividing each tenth into tenths we can reach still further precision and identify the quantity as >24/100 and <25/100. We can, at will, go on subdividing the 'gap' between any pair of integers into any level of detail that we choose.

4.4.3 In this instance I happen to have chosen to subdivide into tenths, hundredths and so on. I might just as readily have chosen to subdivide the unit into halves, quarters, eights, sixteenths, ... or any other kind of part which suited me. The key word is 'suited'; we should choose whichever subdivisions are appropriate to the task in hand and be able to change from one scale to another as we need.

The *values* are all there, only the *names* of the subdivisions vary. We can as readily write or use, as it suits us, ¼, 0.25, 25/100 or 25%; they all mean exactly the same position in the 'gap' between 0 and 1 in our {Real Numbers}.

4.4.4 The gap between any one pair of adjacent integers, be they positive or negative, is exactly the same size as that between any other pair and actually 'contains' the same infinite number of quantities smaller than one whole unit, whatever our chosen scale may be.

4.4.5 The *type of fractions* by which we *describe* those interim values is only a matter of name. All the actual values concerned are the same **rational numbers**.

4.5 Rational numbers

4.5.1 This is actually another of those rather strange labels whose roots lie in the days when only a very few specialists actually studied mathematics and, like all specialists, they spoke their own jargon. What it means is that the value can be expressed as a 'fraction' of the form

$$n/d \quad \text{where both } n \text{ and } d \text{ are integers}$$

4.5.2 I have used n and d only because, in traditional mathematical language, the 'lower' number of a fraction written in this form is the **denominator** and the 'upper' number is the **numerator**.

To **denominate** anything is just to 'give it a name' so the 'lower' number tells us *what kind of subdivisions we have used*. The **numerator**, on the other hand, tells us *how many* of those subdivisions represent the particular quantity we wish to identify.

In this way 1/4, as it is written in computing convention, or ¼ in traditional mathematical notation, tells us that we have subdivided the 'gaps' or 'intervals' between the integers in the real number scale, into *four* equal parts and that our chosen value represents *one* of those parts. In a slightly different notation 0.25 says that we have subdivided the interval into tenths *and* hundredths and that the value = 2/10 + 5/100.

4.5.3 Twenty-five per cent, by another change of notation, indicates that we have subdivided each interval into 100 parts and that our value is denoted by 25 of those parts.

4.5.4 Every 'finite' value less than one whole, i.e. a value whose size or 'magnitude' we can represent in such ways is

$$\in \{\text{Rational Numbers}\}$$

4.5.5 All rationals may be represented in many different ways depending upon the chosen denominator and not only does

$$\begin{array}{ccc}
\{¼ & ½ & ¾\} \\
\downarrow & \downarrow & \downarrow \\
\{0.25 & 0.5 & 0.75\} \\
\downarrow & \downarrow & \downarrow \\
\{25\% & 50\% & 75\%\}
\end{array}$$

but

$$\{\tfrac{1}{4} \qquad \tfrac{1}{2} \qquad \tfrac{3}{4}\}$$
$$\downarrow \qquad\quad \downarrow \qquad\quad \downarrow$$
$$\{3/12 \qquad 6/12 \qquad 9/12\}$$

or any similar mapping with any other denominator.

4.5.6 Remember as you practise such mappings that both numerator and denominator must be integers.

4.6 Irrationals

4.6.1 There are values, which we work with quite commonly, which are outside {rationals} and hence are within

{irrational numbers}

In this case the word 'irrational' does not imply 'silly' or 'eccentric', as it often does in everyday English. All it says is **not** (in the set of) **rationals**.

4.6.2 Examples are all around us and again we must be wary of them in computing since, within a program, they can, if we are not careful, create an endless loop. Not infrequently when a machine has appeared to 'freeze' we have given it an impossible task such as calculating an irrational number!

4.6.3 A very common irrational is the value of π. No matter how long we keep on calculating we never seem to come to an end where n/m can be written in integers, not even at half a million places of decimals have we found a *finite* value for π! Almost equally common are values like $\sqrt{2}$. This too, no matter how diligently calculated, has never yet come to a finite n/m. For π we may, for elementary calculations, use 22/7, or, for greater precision, 3.141915 but neither is a finite, rational, value even if they have that appearance.

4.6.4 {Irrationals} has a very special significance for us since any given value which we may have to calculate, e.g. $\sqrt{2}$; $\sqrt{3}$; $\sqrt{5}$, will not be *identified* as an irrational and we may be unpleasantly surprised by runtime problems with a program which is intended to calculate these.

4.6.5 The second reason is that whenever we call for such a value in a program, if our translator program has something equivalent to 'sqrt(2)', for example, the computer does not 'look up' a table of such values but must calculate them afresh by an internal subroutine. Once again, we need to be aware of the *limitations* of our systems.

4.7 Levels of precision

4.7.1 Some interpreters or compilers confront us with two more number classes, {single-precision reals} and {double-precision reals}. These two, purely computer-related, sets highlight an inherent characteristic of the computer. We may 'juggle' its processor to give increased precision *but* if we want increased precision in calculations we must sacrifice speed. Many computers today use a 'maths co-processor' in order to speed up such calculations and where much numeric work is to be done this is usually a worthwhile investment.

4.7.2 Quite commonly real numbers will be described in the particular computer manual as 'calculated to eight and published to seven significant figures'. Thus a value for π would be calculated as 3.1415926 and shown on the screen, or other output, as 3.141593.

4.7.3 If we need a greater degree of refinement of their value than this we may, by invoking 'double-precision' cause the computer to artificially extend the calculating capacity of the ALU and calculate to 16, publishing to 15, significant figures.

4.7.4 In this case the actual process of computation is considerably slower but the corresponding values for π would be 3.141592653589793, output as 3.14159265358979.

4.7.5 Once again the important thing is for us to realize the implications of very precise calculation of numbers and the cost, in processing time, of doing so. In speed terms {integers} allows the fastest time, {single-precision reals} the next fastest with {double-precision reals} coming a poor third.

4.8 Square numbers

4.8.1 Here are some more mappings; derive the mapping function and name set B in each case

(i)	$A =$	{1,	2,	3,	4,	5,	..., ∞}
		$\downarrow$	$\downarrow$	$\downarrow$	$\downarrow$	$\downarrow$	
	$B =$	{1,	4,	9,	16,	25,	..., ∞}
(ii)	$A =$	{1,	2,	3,	4,	5,	..., ∞}
		$\downarrow$	$\downarrow$	$\downarrow$	$\downarrow$	$\downarrow$	
	$B =$	{1,	8,	27,	64,	625,	..., ∞}

Can you see how example **(i)** relates to:

(iii) $A =$ {1.5,   2.5,   3.5,   4.5,   ..., ∞}
            ↓   ↓   ↓   ↓   ↓
     $B =$ {2.25,   6.25,   12.25,   20.25,   ..., ∞}?

4.8.2 Set **(i)** A is {natural numbers} and
set **(i)** B is {natural square numbers}

which is, it appears another of the 'silly' descriptions since the whole idea of a 'square' is man-made and there are few squares, if any, to be found in nature!

4.8.3 The explanation of this apparent foolishness is that some natural objects can be grouped into squares which have sides whose lengths are natural numbers. It is easiest to see if we use circles, or counters, to represent the natural entities.

In Figure 4.5 the lengths of side of the successive squares is indicated and the growth patterns should be clear. If our natural square number is 25 (i.e. it can be represented as a square containing 25 counters) then its **root** is the length of its side, 5.

By the inverse process a square with sides which are 5 units long will contain 25 counters. In conventional mathematical symbolism, 5^2, implying five multiplied by itself, or 5 * 5, = 25.

4.9 Conclusion

There are many more number sets but for our purposes everything we are likely to meet in everyday computing will probably be in one of those which we have discussed here or in one of their subsets. You should, however, get as much practice as you are able in relating numbers, according to the application, to their relevant set or sets. Such practice should include declaring the variables, defining acceptable ranges and setting up 'go and no-go' test data suitable for use in a real program.

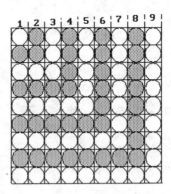

Figure 4.5 Natural squares.

Exercises

(1) Which of the following are {natural numbers}: 4; 4.2; –5; 7; 14.9; $\sqrt{2}$; $\sqrt{4}$; 0.

(2) List six relationships between {natural numbers}, {integers} and {real numbers}.

(3) If, from a group of 20 people eight were male express this as a fraction in three different ways.

(4) Write 2/5 + 3/10 + 7/50 in its simplest form.

(5) Complete the mappings:

$$\{1/16, \quad 1/8, \quad 3/16, \quad ..., \quad 1\}$$
$$\downarrow \qquad \downarrow \qquad \downarrow$$
$$\{0.0625, \quad 0.125, \quad ..., \qquad \qquad \}$$
$$\downarrow \qquad \downarrow \qquad \downarrow$$
$$\{ 6^{1}/_{4}\%, \quad 12^{1}/_{2}\%, \quad ..., \qquad \qquad \}$$

(6) The fraction 22/7 is a traditional approximation of π. Comparing this with the values given in paragraph 4.7.4 to how many places of decimals is it actually correct?

(7) List the members of $\{\sqrt{1}; \sqrt{2}; \sqrt{3}; \sqrt{4}; \sqrt{5}; \sqrt{6}\} \cap$ {irrationals}.

(8) List for $x \le 730$, $\{^{3}\sqrt{x}\} \cap$ {natural numbers}.

(9) What do we mean by 'natural' squares?

(10) We can draw a pictorial model of 3^2 or 3^3. Why can we not draw a similar model of 3^4?

5

A first look at number systems

Objectives

After working through this chapter you should be able to:

❑ understand the column system and the number base
❑ see the relevance of base 2 and base 16
❑ work with modular arithmetic.

5.1 Some number systems

5.1.1 We have four tools for *describing* numbers of all sizes, a fundamental counting group or **number base**; a set of symbols, $\{0, +, -, \cdot, \infty\}$; a set of numerals, in our case, $\{1, 2, 3, 4, 5, 6, 7, 8, 9\}$, which we have already looked at; and a **column system** which allows us to combine numerals and zeros to denote any conceivable number no matter how large or how small.

5.1.2 These things taken together must be counted among the most powerful inventions. The counting group, and hence the set of numerals we use, is determined by the fact that humans happen to have evolved with four fingers and one thumb (digits) on each hand. When early humans (and very young children even today), counted objects they did so by mapping the objects on to their fingers, or putting them into one-to-one correspondence.

The maximum counting group in this case is when all fingers and thumbs have been matched with objects. We then need either another pair of hands or some way of 'noting down' that we have accounted for one 'double handful' of fingers.

5.1.3 The Roman notation reflected this mapping exactly, even putting two 'V's together to make 'X' the symbol for a double handful, as shown in Figure 5.1. It is from such elementary attempts to record and communicate numbers, all over the world and in many different cultures, that the universal number system, in **base 10**, has evolved. We indicate numbers written in this system as, for example, 365_{10}, by adding the *subscript* 10 at the end of the number.

Figure 5.1 The 'finger' mapping of the Roman notation.

5.1.4 The Roman system, however, lacked any indicator for zero and a column system was not developed. As a result the notation was cumbersome and inflexible even though simpler and easier to write than the earlier Egyptian or Greek notations.

5.1.5 Our current numerals are derived from Hindu–Arabic origins and, with the essential contribution of the symbol '0' to represent 'no number' we eventually developed our modern notation. In this the column in which a numeral is written determines *its magnitude relative to the number of units it indicate*. The numeral 5 indicates, for example a group of the size of our earlier group of horses. If the same numeral is written in the next column to the left the number which it indicates is ten times as large – or the equivalent of gathering together into one group *ten* groups of horses of the size of that in Figure 4.1. Put together, the numerals, zero and the columns give us a truly powerful and flexible system.

5.1.6 The concept of position, or column, denoting magnification of a value resembles that of a variety of lenses giving us a variety of 'powers of magnification'. A 'five power', or 5×, lens enlarges objects to five times the size at which they appear to the naked eye; a 10×, or ten-power lens will make things appear ten times as large. We express the 'power' of a position in the columns to 'enlarge' a number as 10 * or 10 * 10 *, and so on. As always in mathematics there is a shorthand for such notations. 10 * 10 * 10 * 10 * 10 * 10 * 10 * 10 * 10 *, as in the leftmost column of the table below, is abbreviated to 10^9 which we say in words as *'ten to the power nine'* or, *'ten to the ninth'* (power).

5.1.7 In this table (which has no end at the left-hand side where the columns can go on for ever and in which the numbers are always ten times as large as in the preceding column), we see various notations, both in symbols and in words. Table 5.1 shows a small part.

Table 5.1

(i)	10^9	10^8	10^7	10^6	10^5	10^4	10^3	10^2	10^1	10^0
(ii)	10*10* 10*10* 10*10* 10*10* 10*	10*10* 10*10* 10*10* 10*10*	10*10* 10*10* 10*10* 10*	10*10* 10*10* 10*10*	10*10* 10*10* 10*	10*10* 10*10*	10*10* 10*	10*10*	10*	1*
(iii)		MILLIONS			THOUSANDS			UNITS		
(iv)	U	H	T	U	H	T	U	H	T	U
(a)			3	0	5	7	2	3	2	7
(b)					4	2	3	5	2	4

5.1.8 Taking the various levels of the table in turn:

(i)   This is the **powers of ten** notation with **no** magnification in the rightmost table, the **units** column, but with each shift along the columns to the left adding another 'ten times' to the degree of 'enlargement' of the number.

(ii) In this section the number of multiplications by ten is detailed. Thus in the 10^6 column, for example, any numeral, let's say 3, represents

$$3 * 10 * 10 * 10 * 10 * 10 * 10 = 3 * 10^6$$

or three million; 3,000,000.

(iii) In the next 'layer' we see the major words for describing large quantities. These are the English subdivisions of numbers and these do vary from country to country.

(iv) In this repetition of H, T, U we see one of the verbal descriptions which most children learn quite early in their lives, 'hundreds', 'tens' and 'units', or, more sensibly, 'units', 'tens' and 'hundreds'. In English verbal descriptions this pattern is repeated so that when combined with level **(iii)** descriptions we continue into 'thousands', 'tens of thousands', 'hundreds of thousands'; 'millions', 'tens of millions', 'hundreds of millions' and so on.

5.1.9 At this stage you should look at rows **(a)** and **(b)** and

(i) both say, in words, and write down in figures, the overall value of each of the two numbers

(ii) write down the value of *each numeral* treating the **units** column as column 1. [For example in **(a)** the 3 in column 3 is worth 300. What is the value of the 3 in **(b)**?]

(iii) compare the sizes of like numerals in different columns by their powers [for example the leftmost 3 in **(a)** $= 3 * 10^9$ and the rightmost 3 is $3 * 10^2$, thus the leftmost is 10^7 times as large as the rightmost. Make *quite sure* that you understand the reasoning here]

(iv) compare the sizes of *unlike* numerals in the same way. Example: the rightmost 2 in **(b)** is $2 * 10^1$ and the leftmost is $4 * 10^5$; the latter is $2 * 10^4$ times as large as the former.

5.1.10 Again, it is important to discuss the principles and practise with plenty of examples to make sure that you are confident with these notations. Showing a small *superscript* numeral to denote the power of a number is known as **index notation** and the examples in **(iii)** and **(iv)**, combining the number of **units** with the **index** is widely known as **scientific notation**. All such notations are of great importance to us in computing where they are commonly written as, for example, 10↑9, since translator programs rarely recognize subscripts or superscripts.

5.2 Other bases

5.2.1 This system should, strictly speaking, be described as the **denary** system but in common language it is usually referred to as the **decimal** system although the decimal notation did not evolve until many generations later. The biological accident of our having ten fingers and thumbs, digits, might just as easily have seen us evolve with eight on each hand!

5.2.2 We should then, equally naturally, have developed numbers in **base sixteen**. Until very recently such thoughts merely gave mathematicians some harmless fun! All that has changed with the introduction of two-state, digital computers (as distinct from 'analogue' computers), in which base 16 numbers are widely used.

5.2.3 Base two numbers need only the numerals 1 and 0 and its column system, essentially exactly the same in nature, sees enlargement by powers of two instead of powers of ten (see Table 5.2).

Table 5.2

(i)	2^9	2^8	2^7	2^6	2^5	2^4	2^3	2^2	2^1	2^0
(ii)	2*2*2*2*2 *2*2*2*2*	2*2*2*2*2* 2*2*2*	2*2*2*2*2* 2*2*	2*2*2*2*2* 2*	2*2*2*2*2*	2*2*2*2*	2*2*2*	2*2*	2*	1*
(iii)	1024's	512's	256's	128's	64's	32's	16's	8's	2's	U
(a)			1	0	0	1	1	1	0	1
(b)					1	1	1	0	0	1

5.2.4 Try to do, with these numbers, the same exercises as before, in section 5.1.8. You may conclude that having words like 'millions' offers some advantages!

5.2.5 We shall need to be fairly fluent in working with the **binary** numbers in practice and need experience to develop confidence.

5.3 Hexadecimal numbers

5.3.1 Describing values in binary numbers, although essential at the very heart of the computer, does take up a great deal of space, especially when we write them or try to manipulate them on paper. For these and many other reasons, but most of all because it is easy to convert to and from base 2, we do actually make a great deal of use of numbers in base 16, or **hexadecimal** numbers, commonly shortened to **hex**. Many pocket calculators today provide for **hex** arithmetic.

5.3.2 Unfortunately we have only nine numerals and zero in our base 10 system whilst we need no fewer than *14* numerals, to accompany zero, in hex. It has become accepted practice, since they are already available both on the keyboard and in computer code systems, to use A_{16} for 10_{10}, B for 11, C for 12, D for 13, E for 14 and F_{16} for 15_{10} and to 'advise' the computer that these are not part of an alphanumeric string, by adding &H before the 'numerals' (since A to F are now also numerals), as in &H1AF, or an 'H' after the number as in 1AFH. (Unfortunately this notation is not standard in computer literature and we must be prepared to work with whatever form is used in the compiler, interpreter or operating system with which we are concerned.)

We do not use the '&H' or 'H' labels in ordinary arithmetic with base 16 numbers but it *is* wise to get into the habit of using the base-indicating subscripts. On the face of it there is nothing to tell us, if we *are* using different bases, what base a number such as 479 may be in although 4F7 *might* be assumed to be in base 16 but we should never rely on such supposition.

In computing, as in mathematics, it is sound practice to assume that if anyone *can* misunderstand or misinterpret what we have written it is inevitable that, sooner or later, someone *will*.

5.3.3 The column system gives us a pattern, shown in Table 5.3.

Table 5.3

	16^9	16^8	16^7	16^6	16^5	16^4	16^3	16^2	16^1	16^0
(i)	16^9	16^8	16^7	16^6	16^5	16^4	16^3	16^2	16^1	16^0
(ii)	16*16*	16*16*	16*16*	16*16*	16*16*	16*16*	16*16*	16*16*	16*	1*
	16*16*	16*16*	16*16*	16*16*	16*16*	16*16*	16*			
	16*16*	16*16*	16*16*	16*16*	16*					
	16*16*	16*16*	16*							
	16*									
(iii)							4096's	256's	16's	1's
(a)			3	A	5	F	2	F	A	7
(b)					4	B	E	E	F	8

5.3.4 Once again you should try all the exercises in 5.1.8 to make sure that you are comfortable with using numbers written in base 16.

5.4 Converting from base to base

5.4.1 Today many computer translators and even many pocket calculators are programmed to convert numbers from one commonly used base to another. It is useful, however, to be able to do so by pencil-and-paper methods and a good start is to make sure that you can list our {natural numbers} in at least our three 'standard' notations.

5.4.2 It is also very useful to look at the addition and multiplication squares for different bases. These, if you have not met them before, are an excellent way of examining the behaviour of numbers and we will look at examples later in the chapter.

5.4.3 To take a listing of the counting, or natural, numbers first. The binary numbers are in 'eight bit' form.

Denary	Binary	Hexadecimal
1	00000001	1
2	00000010	2
3	00000011	3
4	00000100	4
5	00000101	5
6	00000110	6
7	00000111	7
8	00001000	8
9	00001001	9
10	00001010	A
11	00001011	B
12	00001100	C
13	00001101	D
14	00001110	E
15	00001111	F
16	00010000	10
17	00010001	11
18	00010010	12
19	00010011	13
20	00010100	14
21	00010101	15
22	00010110	16
23	00010111	17
24	00011000	18
25	00011001	19
26	00011010	1A
27	00011011	1B
28	00011100	1C
29	00011101	1D
30	00011110	1E
31	00011111	1F
32	00100000	20

5.4.4 When we want to change a base 10 number to, say, base n, we do a repeated division by n and preserve all 'remainders'.

5.4.5 If we wish to convert a base *n* number back to base 10 we do the converse, digit-by-digit multiplication by *n* and include all partial products, i.e. all intermediate results.

5.4.6 For binary conversions the process is a little laborious and we need to take great care not to lose any remainders – especially the zeros!

$$
\begin{array}{r}
2\,|\,\underline{475} \\
2\,|\,\underline{237}\ r\ 1 \\
2\,|\,\underline{118}\ r\ 1 \\
2\,|\,\underline{59}\ r\ 0 \\
2\,|\,\underline{29}\ r\ 1 \\
2\,|\,\underline{14}\ r\ 1 \\
2\,|\,\underline{7}\ r\ 0 \\
2\,|\,\underline{3}\ r\ 1 \\
1\ r\ 1
\end{array}
$$

If we read from the bottom upwards this value is: 111011011 and, by referring back to our BINARY column system we can evaluate this, from the units column, as $(1 + 2 + 8 + 16 + 64 + 128 + 256)_{10} = 475_{10.}$

5.4.7 Once again, if we reverse the process and do the continuous 'multiply by two' routine we see

1	1	1	0	1	1	0	1	1
* 2	2	6	14	28	58	118	236	474
2	3	7	14	29	59	118	237	475
	* 2	* 2	* 2	* 2	* 2	* 2	* 2	
6	14	28	58	118	236	474		

5.4.8 There are several versions of these processes but these are perhaps the simplest. As ever, they need practice to develop confidence.

5.4.9 Conversions from binary to hex and vice versa are somewhat simpler since one **byte**, of eight **bits** (binary **digits**), consists of two **nibbles** of four **bits** each and one hex digit can be represented by four bits. If we look again at our counting numbers, with the binary numbers in 'eight bit' form and separate these bytes into nibbles as shown on the next page:

Denary	Binary	Hexadecimal
1	0000 0001	1
2	0000 0010	2
3	0000 0011	3
4	0000 0100	4
5	0000 0101	5
6	0000 0110	6
7	0000 0111	7
8	0000 1000	8
9	0000 1001	9
10	0000 1010	A
11	0000 1011	B
12	0000 1100	C
13	0000 1101	D
14	0000 1110	E
15	0000 1111	F

We can see that only the right-hand nibble is used for numbers up to 15_{10} or F_{16}, the left-hand nibble, being entirely made up of zeros, is merely space-filling, is worth nothing and may be ignored.

This means that, if we separate out any binary integer, into nibbles, from the units column leftwards, we can transpose each nibble directly into its hexadecimal equivalent.

5.4.10 For 111011011_2 then, we may write this, in nibble form and using 'filler' leading zeroes to complete the leftmost nibble, as

0001	1101	1011,	which we can express as:
1	D	B,	respectively, in base 16, or $1DB_{16}$.

This would give us $1 * 256 + 13 * 16 + 11 = 475_{10}$.

5.5 Simple operations in hexadecimal

5.5.1 For us a very important practical consideration is that in computer manuals all memory allocations are commonly described in hex. In managing memory in PCs we need to identify particular memory addresses and to allocate memory according to the size of file. Finding the hex equivalent of denary quantities and the difference between two hex numbers must, for the modern computer specialist, be a routine skill. If a mouse driver, for example, takes 1460 bytes of main memory we may need to identify the address at which to start loading it and the finishing address at which other programs may be loaded. If it starts loading at &H036F what will be the finishing address? $1460_{10} = 564_{16}$ and $036F_{16} + 564_{16} = 08D3_{16}$. Look back to the work in 5.4.3 and make quite sure that you know why this is.

5.5.2 The use of an **operation** or **table** SQUARE is one way to look at the patterns of number behaviour which arise from the combination of the number system and the laws. Let us, for simplicity's sake, only consider the 'counting up' operations of + and * and the values 1 to 10 only, together with 0, so that we can also see the effects of the laws relating to zero.

To clarify two very important but quite basic ideas by, once again, translating our symbols back into words: 3 + 4 (or any other *a* + *b*), written in full says something like, 'starting at zero count up to three and then count up four more. What is the position reached on the number line?' You should make sure that you can illustrate this, as well as other examples, with a sketch of the process. However, 3 * 4 (or any *a* * *b*), on the other hand, says, 'Start at zero and count up three lots of four. Where is the final position on the number line?' Once again you should be sure that you can illustrate such operations.

Note that wherever we see the + sign we can replace it with the word 'and' with the word 'of' being a similar replacement for the * symbol. Both the symbols are **operators**, sometimes shown as $\oplus$ and $\otimes$, respectively, indicating just what we must do with the two numbers which they indicate.

5.5.3 In this operations square (Table 5.4) we see the complete 'tabulation' of all possible results of any two numbers in base ten, up to ten, being added. Look for patterns and any other features which seem important, discuss them with your colleagues and try to link them with the laws which we discussed in Chapter 1.

Table 5.4

$\oplus$	0	1	2	3	4	5	6	7	8	9	10
0	0	1	2	3	4	5	6	7	8	9	10
1	1	2	3	4	5	6	7	8	9	10	11
2	2	3	4	5	6	7	8	9	10	11	12
3	3	4	5	6	7	8	9	10	11	12	13
4	4	5	6	7	8	9	10	11	12	13	14
5	5	6	7	8	9	10	11	12	13	14	15
6	6	7	8	9	10	11	12	13	14	15	16
7	7	8	9	10	11	12	13	14	15	16	17
8	8	9	10	11	12	13	14	15	16	17	18
9	9	10	11	12	13	14	15	16	17	18	19
10	10	11	12	13	14	15	16	17	18	19	20

When you have looked at this try an addition square for hexadecimal numbers using the facts given in the list of hex numbers in 5.4.3. These relationships are vital to all the work in hex.

5.5.4 If you first try adding the hex numbers 18AF and 2A38, for example, you can add each pair of digits using the list in 5.4.3 and all that remains is to remember to deal with any 'carries'. The answer, 42E7, should then be checked.

5.5.5 By using the same listing of hex numbers you will find that the finding of differences, either by subtraction or by adding complements is exactly the same process as for our ordinary arithmetic. The difference between 28AC and 1689 should be quite straightforward since C – 9 = 'Start at C and count backward to 9' = 'how much must be added to 9 to equal C?'. The answer, 3, can be seen in 5.4.3. Each of the other digits must be dealt with in the same way.

5.5.6 For those subtractions involving 'borrow and pay back' routines we operate exactly as for ordinary, base 10, subtraction. The difference between, for example $4A9F_{16}$ and $37BC_{16}$ is $12E3_{16}$. Make sure that you understand this and that you can work a number of similar examples. It may take a little practice to get used to subtractions where the columns are in 16s rather than 10s but the 'drill' is just the same.

5.6 A system for all real numbers

5.6.1 If we look again at the model in Table 5.1 we may, instead of seeing each leftwards move in the columns as another 'multiply by ten', *start* from the left and perceive that we perform the *inverse* operation in the *inverse* direction.

5.6.2 For each move to the *right* in the columns the value of a numeral is 'diminished by ten times' or is 1/10 the size of its left-hand neighbour. This is shown in Table 5.5 in which it is possible to identify any value smaller than 1 by a mirror image of the integers section of the system.
What is the number at **(a)** now?

5.6.3 As with so many such advances a new symbol had to be devised because we have always used the units column as the reference for the ultimate size of a number and whereas this is clear whilst we keep the full column structure and its labelling, it disappears when we 'rub out' those guides and merely write the number down without them. What became necessary was a foolproof indicator of *the position of the units column* in order to make the value of the number in **(b)** in Table 5.1 clearly different from the number at **(b)** in Table 5.5.

Table 5.5

(i)	10^4	10^3	10^2	10^1	10^0	10^{-1}	10^{-2}	10^{-3}	10^{-4}	10^{-5}
(ii)	10*10*10*10*	10*10*10*	10*10*	10*	1*	10	10÷10	10÷10÷10		
(iii)	U	H	T	U	H	T	U	H	T	U
(a)	·		3	0	5	7	2	3	2	7
(b)					4	2	3	5	2	4

The solution which emerged was to use a dot, at mid-level and to the right of the relevant numeral, to designate the units column for any number. If we write the two numbers using this notation we have, from Table 5.1, 30572327 whereas here we have 305.72327, a very different quantity.

5.6.4 By an extension of the 'each column to the right makes the value one tenth as large' we can denote the new values by *subtracting 1 from the index for each column that we move to the left*. Since $0 - 1 = -1$ the index notation for the tenths column is 10^{-1}, the hundredths column becomes 10^{-2} and so on.

5.6.5 The new system was described as the **decimal** system, meaning tenths-based fractions, as opposed to the **denary** system meaning 'tens based'. As we saw earlier this distinction has become blurred and most computer manuals refer to decimal, rather than denary, values.

What it gave us was an immensely powerful upgrade of our number system so that, as the columns extend *infinitely far in each direction*, we are now able to write *any real number*, no matter how large or how small, in the same notation rather than using two notations as in, for example, 2⅜.

5.6.6 We can and do apply the same principle to the binary system to create a **bicimal** (fraction) system. In this case the columns become ½s, ¼s, ⅛s and so on.

When, in a later chapter, we look at the representation and manipulation of numbers in the computer we shall rely quite heavily on this material so it is well worthwhile developing confidence in our understanding and execution of these ideas.

5.7 A very different number system

5.7.1 First of all look back at Table 5.4 and remind yourself of the 'addition facts' for our ordinary base 10 numbers.

5.7.2 Next check with your own addition square for hex additions. The operations square is clearly a useful tool for exploring any new number system.

5.7.3 Table 5.6 also shows operations on numbers in base 10 and in quite ordinary, everyday, circumstances, but with important differences. Can you offer any explanation of the changes?

Table 5.6

⊕	0	1	2	3	4	5	6	7	8	9	10
0	0	1	2	3	4	5	6	7	8	9	10
1	1	2	3	4	5	6	7	8	9	10	11
2	2	3	4	5	6	7	8	9	10	11	0
3	3	4	5	6	7	8	9	10	11	0	1
4	4	5	6	7	8	9	10	11	0	1	2
5	5	6	7	8	9	10	11	0	1	2	3
6	6	7	8	9	10	11	0	1	2	3	4
7	7	8	9	10	11	0	1	2	3	4	5
8	8	9	10	11	0	1	2	3	4	5	6
9	9	10	11	0	1	2	3	4	5	6	7
10	10	11	0	1	2	3	4	5	6	7	8

5.7.4 Now we look at Table 5.7 showing the first few multiplication operations in base 10. In all the cases we have looked at the numbers could, of

Table 5.7

⊗	0	1	2	3	4	5	6	7	8	9	10
0	0	0	0	0	0	0	0	0	0	0	0
1	0	1	2	3	4	5	6	7	8	9	10
2	0	2	4	6	8	10	12	14	16	18	20
3	0	3	6	9	12	15	18	21	24	27	30
4	0	4	8	12	16	20	24	28	32	36	40
5	0	5	10	15	20	25	30	35	40	45	50
6	0	6	12	18	24	30	36	42	48	54	60
7	0	7	14	21	28	35	42	49	56	63	70
8	0	8	16	24	32	40	48	56	64	72	80
9	0	9	18	27	36	45	54	63	72	81	90
10	0	10	20	30	40	50	60	70	80	90	100

course, go on to infinity in both fractions and negative numbers so we are looking only at a **sample space** of the whole of our number system. What is helpful is that the patterns and the effects of the laws are the same over the whole number system so it is worth examining these tables very carefully to see what they tell us.

Table 5.7 is clearly different from Tables 5.5 and 5.6. What are the reasons for those differences? Are there any similarities; what are they and what do they indicate?

5.7.5 Now let us look at a multiplication square, also in base 10, but which again differs from Table 5.7. Can you offer any suggestions as to why? Table 5.8, shows the same first few multiplications in a base 10 number system, of everyday familiarity, also behaving in ways totally unlike the numbers in Tables 5.5 and 5.7.

Table 5.8

⊗	0	1	2	3	4	5	6	7	8	9	10
0	0	0	0	0	0	0	0	0	0	0	0
1	0	1	2	3	4	5	6	7	8	9	10
2	0	2	4	6	8	10	0	2	4	6	8
3	0	3	6	9	0	3	6	9	0	3	6
4	0	4	8	0	4	8	0	4	8	0	4
5	0	5	10	3	8	1	6	11	4	9	2
6	0	6	0	6	0	6	0	6	0	6	0
7	0	7	2	9	4	11	6	1	8	3	10
8	0	8	4	0	8	4	0	8	4	0	8
9	0	9	6	3	0	9	6	3	0	9	6
10	0	10	8	6	4	2	0	10	8	6	4

5.7.6 Did you get any ideas about the numbers in Table 5.6? Do your conclusions from that investigation hold good for Table 5.8? In case you are still wondering let us write, in words, a couple of very ordinary questions.
(a) If I start work at seven o'clock in the morning and work for nine hours at what time will I finish work?
(b) If school starts at nine o'clock in the morning and we take seven three-quarter hour lessons with a three-quarter hour lunch break when will school end?

5.7.7 We can model **(a)** as 7 + 9 = 4 and **(b)** and **(b)** as 9 + (8 * ¾) = 3 which, compared with our ordinary arithmetic, seems very odd.

5.7.8 The explanation for the oddness of our results, and for those in Tables 5.6 and 5.8 is that the number system of the clock face ends at 12 but also begins again, at the same point, at 0! This means not only that 12 = 0 but that, although we can go on 'going round' the number system for ever, each time we count up to 12 it actually equals 0 and n * 0 = 0. This, in turn, means that the maximum possible quantity which we can express in this system is 11.9. Our ordinary, base 10 system goes on for ever, or is an *infinite* system, because the columns may be extended indefinitely in either direction or into the negative numbers. (See Figure 5.2.)

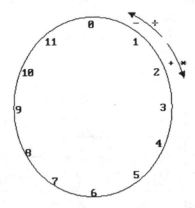

Figure 5.2   Clockface arithmetic.

5.7.9 The numbers in our clockface arithmetic, however, must be less than or equal to 12. This is a **finite set** and, not surprisingly we refer to **finite arithmetics**. One consequence is that $2475_{10} = 3_{mod12}$. If, for any purpose, we need to account for all the 12s which we have 'thrown away' we must do so by another mechanism.

5.8 Finite arithmetics

5.8.1 The size of such a group is also described as its **modulus** (or **mod** for short), and hence we talk of **modular arithmetics**. We may indicate **modular numbers** by a subscript, for instance 3_{mod12} is quite a different number from 3_{10} since it can be the result of an infinite number of different operations on integers.

From the tables above what subsets of {integers} can be combined under addition or multiplication to produce 3_{10}? The normal clockface actually shows two modular systems. Can you identify them? Table 5.8 shows that there are a great many repetitions in the mod12 operations. What happens if we use a prime number modulus?

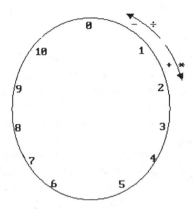

Figure 5.3 A prime number modulus.

5.8.2 An 11-hour clockface is shown in Figure 5.3 and part of its associated multiplication square is in Table 5.9. Complete the square and compare it carefully with the results in Table 5.9.

Table 5.9

⊗	0	1	2	3	4	5	6	7	8	9	10
0	0	0	0	0	0	0	0	0	0	0	0
1	0	1	2	3	4	5	6	7	8	9	10
2	0	2	4	6	8	10	1	3	5	7	9
3	0	3	6	9	1	4	7	10	2	5	8
4	0	4	8	1	5						
5											
6								9			
7				10							
8											
9											
10				8							

5.8.3 Continue the table to column and row 10 and study the results. Discuss your conclusions fully and make sure, by trying other mods, that you have fully grasped the principles of finite arithmetics; they do have many uses in computing. You should also practise converting numbers from their base 10 form to a variety of mods.

5.9 Conclusion

Whilst it is tempting to leave number theories to the mathematicians we are in the very strange position of having to work with a machine which is both very sophisticated and alarmingly simple'minded'. In order to program it at even an elementary level we must have a good working grasp of a the very basics of number systems and number types. We must also be able to function effectively with different number bases and with finite arithmetics.

Above all we must know that it is in the most rule-bound numerical activities that the machine is most error-prone and inefficient unless *we* remain firmly in the driving seat to direct and control its workings.

Exercises

(1) In 473159 how much larger is the 4 than the 1?

(2) What is $3 * 10^3 + 4 * 10^2 + 7 * 10^1 + 6 * 10^0$ worth in total?

(3) What is the value of $(3 * 10^7 - 4 * 10^5)$?

(4) Convert to base 16:

    (a) 439_{10}

    (b) $16,000_{10}$

    (c) $32,160_{10}$

    (d) 3974_{10}.

(5) Convert to base 10:

    (a) $4AF_{16}$

    (b) 110011101_2

    (c) $FFFF_{16}$

    (d) $ABCDEF_{16}$.

(6) Evaluate:

    (a) $349 * 10^{-3}$

    (b) $0.0285 * 10^6$

    (c) $349A * 16^{-2}$

    (d) $2.AF3 * 16^3$.

(7) Evaluate:

    (a) $(11 * 4)_{mod12}$

    (b) $(11 + 3 + 7 + 9)_{mod12}$

    (c) $(7 * 10)_{mod11}$

    (d) $((6 * 9) + (5 * 8)+(9 * 7))_{mod11}$.

(8) Evaluate in mod12 the following values in base 10:

    (a) 43

    (b) 179

    (c) $(25 + 14 + 17 + 13)$

    (d) $4 * 3 * 7$.

(9) Evaluate $y = (2x + 3)_{mod11}$ for $1 \leq x \leq 10 : x =$ integer.

(10) Find all members of the set $R = \{x: x$ integer $0 \leq x \leq 4\}$ and $4x = 2_{mod5}$.

6
Mistakes and errors

Objectives

After working through this chapter you should be able to:

❑ distinguish between mistakes and error
❑ recognize how and where mistakes are likely to be made and understand elementary detection procedures
❑ understand inherent and induced error, the absolute and relative size of error and how error increases in processing
❑ identify error induction in the computer
❑ calculate error by commonly used methods
❑ recognize how error is propagated and take elementary precautions to minimize error.

6.1 Mistakes

6.1.1 One essential difference which we must always have in mind when adapting mathematical ideas and techniques for practical use in computing is that between **mistakes** and **error**. We have, in everyday speech, somehow managed to diminish the word 'error' by using it as a 'softer' criticism. Rather than simply saying, 'You have made a mistake', we have persuaded ourselves that it is less offensive to say 'I think that there is a small error here'.

We also often even refuse to admit, at least to anyone else, that we ourselves ever *make* mistakes!

6.1.2 In the much more disciplined world of computing, as in mathematics, it is important to realize that, although all human beings make mistakes and may, if they are sensible people, learn from them, *any single mistake could have been avoided.* It is also very important to learn to admit to our own mistakes and not try to 'bury' them!

6.1.3 All our systems should, as a matter of common sense and good elementary design, take account of these natural human failures. To ignore this in the hope that it will not affect *our* systems is simply foolish. We *must* assume that mistakes *will* always be made, try to help people detect and correct them and, most important of all, ensure that a system will not be put in jeopardy by any mistake which we might reasonably have foreseen.

We must also encourage the best possible training for the people who work with our systems, encourage them to seek help when a mistake *has* been made and not to be afraid to 'own up'. Considering the garbage in–garbage out (GIGO) problem in computing data-prep(aration) staff are often under-trained and vastly under-rated. If a mistake has been made, at data-prep, data-input or in processing, the output *will be corrupted*, and, worst of all, this will not necessarily be apparent!

Today this problem has become much more serious because, with networked PCs, many users have no data-quality control training at all. Our systems must be even more carefully protected.

6.1.4 Mathematics can do nothing to prevent mistakes being made; it can, however, offer some elementary help in *detecting* mistakes *before* they can affect the integrity of the processing or cause degradation of the output.

Mistakes in processing will almost always be either as a result of imperfect program design or testing *or* in some failure on the part of the operator. However, the most vulnerable of all the elements of any system is the man–machine input interface.

6.1.5 Among the kinds of mistake which are commonly made in any data input phase are:

❏ mislaying source documents
❏ misreading of data
❏ wrong data type
❏ wrong data range
❏ miskeying data
❏ transposition of characters or numerals
❏ character drop-out
❏ bit drop-out in data transmission from remote input.

The list is by no means exhaustive but will serve to indicate which types of mistake may be detected by *some* mathematical process. We have talked earlier about validating data types and data ranges which means that well designed *and tested* data-input validation routines are elementary requirements.

6.1.6 The matter of *range* is especially important for two reasons. If, for example, any part of a date is out of range it should be detected as part of the validation/verification routine. It ought not, for instance, to be possible for a computer system to issue voter registration papers for a five-year-old child. Even more important is that any value *which is highly improbable* when range tested should trigger an *exception report* and lead to action. This process may well bring together the statistical models and techniques and the range testing subroutines. It should certainly prevent an ordinary domestic consumer being sent a ridiculously high electricity bill, for example. This could be avoided by referring for checking any account which was more than three times the standard deviation above the mean for domestic consumers. Bear this sort of testing in mind when you work on Chapter 12 later on.

6.1.7 Batch quantities and batch totals, counting and summing relevant numbers of items (for example, money fields), in a set of documents, are fairly obvious examples of 'ordinary' arithmetic being used to detect mistakes.

A little less orthodox is the idea of the 'hash' total. We can, for verification and validation processes *only*, add together any set of numeric fields, invoice dates for instance, or customer numbers, to generate check totals from ordinal numbers. Indeed these are so useful for detecting mistakes that it is often worth using purely *numeric* data types for such purposes even if alphanumerics have been used in the preceding manual system. Hash totals may equally well be used, internally, for *alphanumeric* fields of special importance.

Since all the letters and symbols are dealt with in the machine by numeric codes we may total the codes rather than the symbols which they represent. All the character codes in, say, a sequence of characters in transmission or even of all the code of a program, may be added to form a 'checksum'.

The use of parity bits in checking for any corruption of a single character is almost too well known to merit lengthy discussion. In **odd parity** a 1 is placed in the 'leftmost' bit position of an eight-bit byte when there is an *odd* number of binary 1s in the remaining seven bit positions. If 'A', coded in ASCII seven-bit code as 100 0001 is represented in odd parity it becomes 0100 0001 because it does not have an odd number of binary 1s in the original seven-bit form. If it were corrupted into, say 0100 1001, at some point a parity check would reveal that something was wrong since with three 1s in the rightmost seven bits the **parity bit** ought to be 1. Clearly this can only be used, on an eight-bit bus or communications channel, if we need only seven bits to code all the characters which we use. Since today we often need the extended, eight-bit character set we may be restricted to the 'no parity' condition. Parity checks cannot then be used.

6.1.8 One other very common group of mistakes is those leading to copying in ordinary numbers to a data-prep form or the keyboard.

It is extremely easy, between what our eyes see and our fingers input, for 1058535, say, to appear as 105535 or 1085535. Each of the three forms is the correct data type, within a similar data range, and hence unlikely to be detected by the simpler validity checks. Such mistakes are, however, among the most frequent in any data-handling and transcription process.

If it is a number of some importance which is in frequent or permanent use, such as a customer number, or even if it is used as an internal hash totalling process, we may use a standard arithmetic process for generating a hash total and appending it to the end of the number as a checksum.

With 1058535 we could add the numerals together, as $1 + 0 + 5 + 8 + 5 + 3 + 5$ to give a total of 27 which we could then tack on to the end of the number to make 105853527.

This, on the face of it, *would* show up a digit drop-out, *if* we knew where the original number 'stopped' and the checksum 'began'. We don't! Furthermore we have no *way* of knowing since the checksum may, in this type of example, be any size, almost at random. For instance the checksum for 23 would be 5; for 28 it would be 10; how would we know how big a checksum to expect or where to 'cut it off'?

6.1.9 This is a natural feature of any **infinite** arithmetic; the numbers just keep on growing. Only a **finite** arithmetic will give us numbers of fixed length, no matter how large the 'original' number. If we used mod7 numbers, for example, the checksum, after adding all the digits, for 1058535 would be 6, for 23 it would be 5, for 28 it would be 3; always a single digit to append to the subject number and to strip off for checking.

Unfortunately, however, while this *would* detect any digit drop-out (unless that digit were 0, which is not unimportant), it would *not* detect a transposition of digits, as in 1085535, because the checksum would be the same, and in consequence, so would be the check digit.

It is to overcome this problem that we do some more unorthodox arithmetic by 'weighting' the columns before we hash total the digits. We do this by multiplying the numeral in each column by a *different* fixed factor. In ordinary commercial practice we keep it as simple as possible by giving 'weighting factors', that is to say multipliers, to the columns in the sequence 2* for the units, 3* for the tens, 4* for the hundreds, 5* for the thousands and so on. In our example we should then have

$$8*1 + 7*0 + 6*5 + 5*8 + 4*5 + 3*3 + 2*5 = 117_{10}$$
$$\text{and } 117_{10} = 5_{\text{mod7}}$$

If we now check our transposition error, where we had 1085535 our checksum would be $120_{10} = 1_{mod7}$ and we should know at once that the number had been incorrectly entered.

6.1.10 In practice the weightings may be kept secret, or even, where security is of exceptional importance, be changed at frequent intervals.

The check digit actually used is customarily the **complement** of the **modulus** as this leads to a considerable improvement in the simplicity of checking. Applying this to our first example, using mod7, for instance, the next to the last step would be to *subtract* the modular value of the checksum from the modulus itself, 7 – 5, to generate the check digit 2. When we have taken this final step we can make 1* the weighting factor of the new 'units' column, containing the check digit, and recalculate the checksum without 'stripping off' the check digit. If all is well the new checksum should be 0, whatever the modulus.

Let us verify that with our number, 10585352 and weight, from right to left, by factors of 1, 2, 3, 4, 5, 6, 7 and 8, respectively. You should arrive at a checksum of 119_{10} which, in mod7, equals zero and shows that the number has probably not been corrupted. I use the word 'probably' because there are some transpositions which will not be detected. Can you suggest why? Some 90% of such mistakes *will be detected* and that makes the technique very important to us.

6.1.11 Before leaving this topic we should look back to the previous chapter and our exploration of finite arithmetics. We saw that the prime number 11 had many fewer repetitions in its number squares than did 12, when used as a modulus. For this reason we always use a prime number modulus, often mod11. What we need then, however, is another symbol, commonly X for 11 itself (with 0 for 10), if we are to keep to a single check digit.

There are advantages to using still larger numbers, but the small increase in mistake-trapping is usually outweighed by the additional complications in processing. There is always a very difficult judgement to make as to when relatively small gains, in effectiveness or efficiency in one aspect of an application, are at the expense of slowing processing to an unacceptable degree. We can take it that mod11 is probably the commonest in commercial and industrial use.

6.2 Error, error propagation and control

6.2.1 To the 'pure' mathematician numbers are clean, precise, even beautiful. In the real world, in which we have to function, pure numbers are of little importance and numbers are *usually* imprecise, approximate or may even be downright treacherous! Any real life number, 3, for instance, must always be accompanied by some label. We need to know 'Three what; ..., dogs, metres, dollars, ...?'. The number is useless to us unless it is *describing some reality*. We talk then of *applied* mathematics, i.e. applied to the description of real entities and solving real problems.

6.2.2 What *we also have to be concerned about* is that our machine will only process the 'pure' numbers. We must strip off the labels *before* processing and ensure that we replace them afterwards because a number on its own makes no sense to people in the real world. This is a primary purpose of our variable names; they must be mapped on to the labels of our numbers.

6.2.3 Whenever we label a number, however, it is because it is describing a measurement on some **scale** and this is where the world of **error** takes over from the neat and tidy world of pure mathematics. Every number which is a model of reality has, with very few exceptions, **inherent error**. Only when we are using {natural numbers} for *counting a small collection* of entities, using the positive integer number line as our scale, do we have numbers without this error, which is *not* artificial and *not* due to mistakes or carelessness. Nor can it be removed, no matter how concerned we may be with accuracy. It is an entirely inevitable attribute of the number in its scale.

6.2.4 There *are* situations in which a number will be rendered less precise because of something which *we* do, directly in calculations, indirectly in the way we write our programs *or because of the limitations of the computer itself.* We call the effects of these, by contrast, **induced error**, because they have been 'brought into' the situation, as it were from 'outside' the numbers themselves. Error will grow during processing: unless we are extremely careful in our computer processing it may well grow out of control. Every inherent error may be looked on as a seed from which much greater error may grow and we call that process **error propagation**. There is, as you see, a considerable new vocabulary to be learned and, a little later, some new symbols.

6.2.5 First let us look in more detail at inherent error. To take a rather crude illustration first of all we can see, in Figure 6.1, the effect of measuring a value, *V*, shown by the thick line at the top, on a scale of units where, in this instance, we can see that $2 \leq V \leq 3$, or *V* is somewhere between 2 and 3. If we estimate,

Figure 6.1   A number on a unit scale.

by eye (see Figure 6.2) we can see that $2\frac{1}{2} \leq V \leq 3$, but V is still 'somewhere in between' and we can never be quite sure just where.

Figure 6.2   'Eyeing' the halves.

6.2.6 There is always this element of uncertainty and it is in this that our concept of inherent error originates since any value lies in just such an area of vagueness and we can never be quite sure where.

6.2.7 No matter how much we may subdivide our scale into smaller units there is still a region of uncertainty, one unit in size. Figure 6.3 gives an illustration of this and the consequent inevitable nature of inherent error.

Figure 6.3   There is always uncertainty within one unit (or scale division).

6.2.8 When we state that a measurement has the value 3 we are saying that, on the scale used, its value is greater than $2\frac{1}{2}$ and less than $3\frac{1}{2}$ *but we don't know, and cannot know, just where, between those boundaries the 'true' value lies!* To indicate this uncertainty we should, if we were being extra careful, write that the value is $3\pm\frac{1}{2}$. Ultimately, no matter how fine the subdivisions of our scale, the **error bound** is always $\pm\frac{1}{2}$ of one of such subdivisions.

6.2.9 In all cases such as these the numbers have a positive and a negative error component of approximately the same size. This means that the error bound is always symmetrical and bilateral, that is to say that it always has *two* parts of equal size. In our example the total error range, within the error bound, is one unit, but the largest amount by which any *actual value* may differ from the **reported value**, 3, (and ignoring the ± symmetry), is $\frac{1}{2}$.

6.2.10 When we ignore the sign of a value and only concern ourselves with its magnitude we speak of the **absolute value** of that magnitude which we write as $|\frac{1}{2}|$ or abs($\frac{1}{2}$), we may find this expressed in symbols as $E_a = 0.5$.

6.2.11 Here we must take note of the ultimate case of the effect of E_a. The difference between any **two** values **reported** as 3 is that one of them could be as small as 2.5, the other as large as 3.5. In other words, when we are comparing two or more values, reported as identical, the possible error is $2 * E_a$!

This is of the greatest importance in computerized manufacture, or any other application of computing to any field of engineering. If a cylinder is bored to a diameter of 30 mm and the piston is turned to the same reported (or nominal) size it could happen that the piston was 30.5 mm in diameter and the cylinder 29.5 mm in diameter! Not much hope of a good fit there. Absolute error, therefore, may present some unexpected difficulties if we are not on guard against such extremes.

6.2.12 In manufacturing industry it is assumed that sizes of any reported value may fall anywhere within the error bound and will be distributed statistically. Process control will often call for computerized statistical models to suit.

6.3 Relative error

6.3.1 Clearly, if all measurements on the same scale have the same error bound and consequently the same absolute error the *effect* will differ according to the reported value. $E_a = \frac{1}{2}$ on a reported value of 3 will be comparatively large whilst $E_a = \frac{1}{2}$ on a reported value of 50 will be much less so. For this reason we find the absolute error useful mainly as a stepping stone towards the more meaningful **relative error**, commonly written as E_r. In this we express the absolute error *as a proportion of the reported value* and, as with many other proportions, we usually express it as a percentage. (We can also convert the error bound to a *relative* percentage to give us two possible forms.)

(a) The expression $100 * (E_a/R)$ (where R is the reported value) gives the *absolute* value of E_r, in the case of our reported value of 3, then, we have $100 * (\frac{1}{2}/3) = 16.6\% = E_r$; *or* any *individual* value reported as 3 may be 16.6% larger or smaller than the reported value.

(b) The expression $100 * (\text{error bound}/R)$ is maximum possible relative error between any two values reported as R.

In our example once again, $100 * (1/3) = 33.3\%$, is the maximum possible E_r between any two such values. We most commonly use method **(a)** but we must be aware that if we happen to forget the 'larger or smaller' proviso we could seriously underestimate E_r in some situations.

6.3.2 If we now compare our reported values of 3 and 50 using E_{r1} for the former, at 16.67% (corrected to two places of decimals), how does this compare?

$$100 * (\tfrac{1}{2}/50) = 1\% = E_{r2}$$

This makes it very clear that the *size* of any reported value is a very important factor in any consideration of its inherent error. But only up to a point where any practical 'fit' is concerned. A piston of 50.5 mm diameter will still *not* fit into a cylinder of 49.5 mm diameter!

6.4 Error propagation

6.4.1 Let us, at this point, make our first investigation of **error propagation**.

If we take reported values of 2 and 3 units and add them we have a reported result, 5. Taking account of inherent error we have, at the minimum

$$1\tfrac{1}{2} + 2\tfrac{1}{2} = 4 \text{ and, at the maximum}$$

$$2\tfrac{1}{2} + 3\tfrac{1}{2} = 6 \text{ and } E_a = 100 * (1/5) = 20\%$$

E_a on the 2 is 25% and on the 3 is 16.67%.

6.4.2 On adding two values the final, or accumulated, error is somewhere between those on the individual values. Can you, by taking a number of such examples, derive any rule for the accumulation or propagation of error under addition?

6.4.3 To examine the effects in multiplication we may use a visual model of the situation such as $A = L * B$. In Figure 6.4 we see the rectangle, in thick outline, whose reported sides are 2 and 3 units, respectively.

The reported area is six square units. The smallest possible rectangle to fit the 2 * 3 reported lengths is shown hatched and is $1\tfrac{1}{2} * 2\tfrac{1}{2}$ or $3\tfrac{3}{4}$ square units; the largest is $2\tfrac{1}{2} * 3\tfrac{1}{2}$ or $8\tfrac{3}{4}$ square units. The error bound is $-2\tfrac{1}{4}$ to $+2\tfrac{3}{4}$

Figure 6.4 Multiplication explored as $L * B$.

square units and we see that **induced error** may not be symmetrical. This, in the long run, is quite important but let us, for the moment take E_a to be half the error bound which makes the error 2½ square units! Note that we still need the label for our absolute error although it disappears when we write E_r since that value is always a proportion. In this case E_r is clearly 41.67% since the model 100 * (E_a/R) produces that answer. In other words we have an apparent error of 41.67% as a result of multiplying quantities with E_r of 25% and 16.67%, respectively.

If we take the larger 'side' of the error bound as the safe margin, E_a would be 2.75 rather than our 'nominal' 2.5 and E_r would be 45.83%. Clearly we can say that induced error in multiplication is found by summing, adding together, the contributory inherent errors, if the error bound is symmetrical.

If, as is often the case, the error bound is not symmetrical the induced error may be slightly more or slightly less than the nominal error.

6.4.4 Since, in computing, we often set up our models to perform whole sequences of multiplications let us take one step further and look at three-stage multiplication by way of modelling a cube of sides 3 units long.

The reported value of the volume will be, of course, 3 * 3 * 3 = 27 cubic units. The smallest, shown hatched in Figure 6.5, will be

$$2.5 * 2.5 * 2.5 = 15.625 \text{ cubic units}$$

with the largest 3.5 * 3.5 * 3.5 = 42.875 cubic units. The error bound is, in total 27.25 cubic units, which is *larger than the reported value itself!*

For any practical purpose the result is useless although it is a dramatic illustration of the propagation of error in deceptively simple models. If we quantify it as E_r we find that it is again non-symmetrical but the average of the negative and positive elements is 13.625 and E_r is 50.46%; this is three times the inherent error in the length of a side.

Figure 6.5 $N * N * N$ in the form of a cube.

6.4.5 You should try several examples of such additions and multiplications to be quite sure that you understand these effects. They are of the greatest possible importance to us.

6.4.6 If we look at subtraction, sticking to our reported 2 and 3, 3 − 2 can be, at the most 3.5 − 1.5 = 2 ; it may be, at the least, 2.5 − 2.5 = 0 giving E_a = 1 and E_r = 100%. Try some other examples and see if you can arrive at a rule for error propagation in subtraction.

6.4.7 In division 3/2 has a maximum value of 3.5/1.5 = 2.3˙ and a minimum of 1: E_r is then 44.4%. You should try a number of other divisions, starting with the inverse, 2/3, and see if you can evolve a general rule for division.

6.4.8 Finally try

$$\frac{2 * 3 * 4}{5 * 6 * 7}$$

(a) first multiplying all the top row, then all the bottom row and doing the division
(b) second doing a multiplication followed by a division and so on
(c) third 'cancelling' any possible pairs first and then solving the rest.

Which process causes the least error propagation and how does this affect program design?

6.4.9 The problem that we have in practice is three-fold:

(a) to recognise how and where error occurs and is propagated
(b) to control and limit that propagation by careful program design and testing;
(c) to present our output values with due regard for their error component.

6.5 Rounding numbers

6.5.1 There are other important sources of induced error in our ordinary arithmetic processes, whether or not we use a computer. We often ask for values to be 'rounded to one or more decimal places'. This is usually done to simplify manual arithmetic and is less common with the use of pocket calculators – except that some of these offer options to round to n places.

6.5.2 The most familiar process is to designate by a vertical bar (|) or something similar, the column at which rounding is to take place, for example

(a) 3.141|1592

(b) 3.1415|92
(c) 4.743|5
(d) 4.744|5

and if we describe this as the 'cut' we may say, 'if the numeral in the column immediately to the right of the cut is >5 add one to the numeral in the column immediately to the left of the cut; if the numeral to the right is <5 do nothing; if the numeral to the right is 5 then *either*, if that to the left is odd do nothing, *or* if that to the left is even add 1.'

Perhaps we should pause for a moment here to note that when we round *up* we actually make a number larger than it was originally, if we 'do nothing' we are actually cutting off the remaining digits and making the number smaller. By taking the 'odds or evens' approach to the right hand 5 we are balancing out the ups and downs. The error bound induced is then ± 0.5 *of the value of the 'cut' column*. If we cut at the third place $E_a = 0.0005$ and when we look at any number which has been corrected to, say, three places of decimals, commonly written 3D, this is the error which has been induced into it.

To return to our examples, the 'new' numbers above are:

(a′) 3.141
(b′) 3.1416
(c′) 4.743
(d′) 4.745.

If we were simply shown these numbers, in ignorance of their *original size*, we would have to apply $E_a = 0.0005$ but here we *do* know what they were originally and we can compare the *actual* with the *rounded* value. If, for example, we look at that difference, and we find differences by subtraction, ignoring the sign if we need the **absolute difference**, in the case of example **(a)**, above, $|a - a'|$ is 0.0001592 and this is the E_a which we have induced by rounding to 3D. The resulting error is *unilateral* and E_r is 0.49% which is quite small. (How long, however, would it remain small if it were involved in a sequence of multiplications, say?)

Notice that in this case the error bound is unilateral, one-sided. Try the others in that little group and arrive each time at E_r. Then see what the effect would be if we rounded to 2D.

6.5.3 This type of induced error is one which we can control relatively easily by careful consideration of the processes involved and by postponing all rounding to the last possible moment in our calculations. It is a very good argument for using the high precision variables in many commercial calculations; subject always to the cost in processing speed.

6.6 A rounding model

6.6.1 It occurs quite often that we, for various purposes but especially for the meaningful rounding of numbers *before* publication, need to include a rounding procedure within a program or a spreadsheet.

What do we need to do?

(1) decide a 'cut' point
(2) decide whether we should round up or down
(3) achieve the rounding.

Here is a possible pseudo-code model for rounding to 2D

```
read n
A ← INT(n*1000)
B ← INT(A/10)
IF (B−((INT(B/10))*10))/2 = 0
    OR IF A−(10*B) > 5
        THEN A ← (A+5)
    ENDIF
n ← (INT(A/10)/100
write n
END
```

This is not necessarily a very good model, but you should test it thoroughly and try modifying it for rounding to 3D for the value of the practice. It does show the decisions which we make in our minds when we are rounding numbers and it shows some of the ways in which we can 'tell' the computer where our cut comes and how to deal with the 'next column to the right, next column to the left' part of the decision making. The machine, of course, doesn't *know* right from left.

There may be, in the translator for the programming language which you are using, 'invisible' subroutines, called by a single command, such as **cut** or **fix**, which will accomplish part or all of such a task. Even if they are present we must still understand what they do and how they do it if we are not to be blind when dealing with error propagation.

6.6.2 In a situation where the exact balancing of the 'rounding of the fives' is not too critical we can, in a spreadsheet for example, often use a single formula such as (INT((n * 1000) + 0.5))/1000 to round to 3D.

6.7 Significant figures

6.7.1 A further source of induced error is the reduction of numbers to significant figures. The argument goes something like this: if we have a number of substantial size, say, 1058535792, the last few numerals, say the 5792, represent a part of such small size compared to the overall size of the number that they are really insignificant. Indeed they may well be completely spurious as in stating that the population of Ruritania is 53475372 since no-one, no matter how accurate the counting process, could arrive at a number of such precision. Where such numbers give a false impression or where they are too cumbersome in our arithmetic, let us quote only those five figures, say, which give an adequate impression of size.

The result is that, to five significant figures (often written as 5 sig. figs), 1058535792 would become 1058500000 and 53475372 would be 53475000. It is clear that we have created or induced error in just the same way as by rounding to nD. Again the error will be unilateral and again it may not have a disastrous effect on the original number. The revised value, however, must only be used in computations if the consequent error propagation has been foreseen and guarded against.

6.7.2 We can, of course, express real numbers to n sig. figs; in 2.0375612, just as in our population count, above, the last few digits are very small compared with the left hand 2. We could well express this as 2.0375, i.e. to 5 sig. figs. Note that any zeros essential to the true size of the number must be kept.

6.8 Truncation – a procrustean bed!

6.8.1 In ancient Greece a villainous man, Procrustes, would entertain guests generously so long as they would consent to spend the night under his roof. They had, however, to agree to sleep on his 'guest bed'. Once there, alas, if they were too short to fit the bed they were stretched, forcibly, until they did; if they were too long – the 'extra' bits were chopped off!

6.8.2 Modern computers also behave like this because register size is the first of another set of error-inducing situations within the ALU. Each register has a finite maximum size of number which it can contain and numbers with digits in excess of this will be subject to induced error in two different ways.

By **chop** or **cut** where any excess digits at the right-hand end of a number, i.e. the least significant digits, will simply be lost. This induces 'normal' rounding or sig. figs measures of error and can be accounted for comparatively easily, *so long as we are prepared for it.*

6.9 Overflow error

6.9.1 Much worse is **overflow** error, where it is the leftmost, **most significant**, digits which are discarded. This, of course, may be quite catastrophic and is usually irrecoverable. Fortunately many writers of compilers and interpreters help us by delivering an **overflow error** message but the cut may already have been made even though our processing has been halted. The prevention of, warnings about and recovery from, overflow errors is critical.

6.10   Changes of number base

6.10.1 A final example of internal error-induction is in the **conversion errors** induced when changing decimal fractions to bicimal form or {binary fractions} = {1/2, 1/4, 1/8, 1/16, ...}. If we take any decimal fraction which is a sum of any of the elements of this set we have no problem. For example, 0.75 is fine because it can be expressed also as 1/2 + 1/4, but let us look at 0.7, which would, on the face of it, seem a simpler value to transform. For convenience let us arrange for the conversion to follow a similar form to the denary/binary conversion we looked at earlier but we will put the binary result on the top line rather than at the side, as in 'long' division. We will, at each level divide by the appropriate binary fraction in the sequence but, for consistency of number operations, express these in decimal form.

6.10.2 First, how many halves in 0.7; how many quarters in the residue; how many eighths; how many sixteenths and so on.

$$
\begin{array}{r}
\underline{0.10110011}\ \text{Base-two fraction} \\
0.5\,|\,\underline{0.7} \\
|\,\underline{0.5} \\
0.25\,|\,\underline{0.2} \\
0.125\,|\,\underline{0.200} \\
|\,\underline{0.125} \\
0.0625\,|\,\underline{0.0750} \\
|\,\underline{0.0625} \\
0.03125\,|\,\underline{0.01250} \\
0.015625\,|\,\underline{0.0125000} \\
0.0078125\,|\,\underline{0.0078125} \\
0.00390625\,|\,\underline{0.00468750} \\
0.00195325\,|\,\underline{0.00390625}\ \text{and so on, ...}
\end{array}
$$

6.10.3 This would be comparatively straightforward to program and you should try to write a pseudo-code version.

There is an alternative model which is perhaps not so easy to understand but is much simpler to set out for pencil and paper working. In this we 'fix' the columns firmly on the decimal point and perform repeated multiplications by 2 of the digit(s) to the *right* of the decimal point.

As the first step

$$0.7 * 2 = 1.4 \text{ and the '1' is the first digit of our answer.}$$

Further digits will be all the ones and zeros as we go along. The second step, multiplying only the digit to the right of the decimal point, will be

$$0.7$$
$$1.4 * 2$$
$$= 0.8$$

and the complete process, to 10 'places of bicimals' will be:

Read down ↓

```
 |.7 *2
1|.4 *2
0|.8 *2
1|.6 *2
1|.2 *2
0|.4 *2
0|.8 *2
1|.6 *2
1|.2 *2
0|.4 *2
0|.8 *2
```

Take it several stages further and see if the '0011' sequence continues, or whether the bicimal fraction 'works out exactly'.

If there were room only for six of the bits in the above answer, which is, of course, the same by either method, can you see how to evaluate the consequent error?

As always you should try this for a variety of numbers to be sure that you have grasped the routine.

6.10.4 It is not difficult to see that there will be many such finite decimal values which generate irrationals in the binary system and will be truncated when the register is filled. This used to be something of a problem in the early pocket calculators which baffled many people when (10/3) * 3 did *not* return the expected answer! The programmers of those processors now look after us a little better.

6.11 The control of error

6.11.1 As an example of the need for the control of error in processing let us look at the model of the volume of a cylindrical cone:

$$V = \tfrac{1}{3} \pi r^2 h$$

where r is the radius and h is the height. If we assign some test values: $r = 2.2$ cm and $h = 12.5$ cm and take π as being 3.141 we have:

$$V = 0.3 * 3.141 * 2.2 * 12.5$$

and we must next look at the error in each number or process in order to assess the extent of error propagation and whether we can minimize it.

We must also do this, as a routine part of our modelling/programming so that we can establish, and this is really extremely important, *what degree of accuracy we may legitimately claim when we publish our answer.* We far too often publish results to many more places of decimals, or to many more significant figures that can possibly be justified by the accuracy of our numbers or of the processing! Such answers are not merely pretentious but are often dangerously misleading.

6.11.2 Let us subject our model to such a scrutiny. The errors present are:

n	E_a	$E_r\%$	
0.3	0.05	1.67	induced by rounding
3.141	0.0005	0.02	ditto
2.2	0.05	2.3	inherent – but more will be induced by conversion (below)
12.5	0.05	0.4	inherent.

This totals, since all elements are to be multiplied, 4.4%. This means that a calculated value of 25.91325 we have an error of ±1.14 cubic centimetres. How meaningful are all those decimal places in our uncorrected answer?

6.11.3 The next step is to see if we can, by planning our calculation, reduce that error. What happens if, for example, we work in integers and do all our reconversion at the last stage. This would imply also dividing by 3 at a late stage rather than multiplying by ½.

Try this and compare the two results. This is just an illustration of how we must examine all our operations on models to exercise control over error propagation and to publish only reliable results.

6.12 Conclusion

Given all these ways in which error is present in numbers, is induced in our operations and propagated by our processing it is small wonder that any computer person who is without a firm grasp of these matters may experience some very severe shocks and allow some very strange results to be published.

Exercises

(1) List five types of mistake which may occur during the data prep and data entry stages of the processing cycle.

(2) Give two examples each of the appropriate use of hash and batch totals.

(3) Describe the principle of a parity check and how even and odd parity differ.

(4) Using mod7 and weightings of 2, 3, 4, 5, ... for the units, tens, hundreds columns, and so on, append appropriate check digits to:

    (a) 439

    (b) 27946

    (c) 30000

    (d) 1058535

    (e) 17493726.

(5) Using the same procedure as in question (4) determine which of the following numbers, which include check digits, are valid:

    (a) 23756

    (b) 1794

    (c) 200004

    (d) 1743562

    (e) 10545722.

(6) (a) Is there any number in question (5) in which there could have been a digit transposition which the check-digit procedure would not have revealed?

    (b) How could this happen?

(7) What is the error bound of values reported as:

    (a) 6 km

    (b) 3.5 cm

    (c) 1.05 grammes

    (d) 350 cm?

(8) Calculate the maximum relative error on:

    (a) 3.5

    (b) 13.5

    (c) 130.5

    (d) 1300.5.

(9) Calculate (3.5 * 13.5 * 1.5)/(7.0 * 4.5 * 40.5) in such ways as will:

    (a) maximize

    (b) minimize, the accumulated error.

(10) What is the effect on error if 3.5 * 7.25:

    (a) calculated to integer precision

    (b) cut to 1D

    (c) rounded to 1D

    (d) calculated to 3 sig. figs

(e) worked out 'in full', with no cut or rounding?

(11) If $V = \frac{4}{3}\pi r^3$ is to be calculated to 3D what precautions, if any, should we take about the values used in the calculation and the sequence of calculations in our model?

(12) If, correct to 2D, $x = 2.4$ and $y = 1.9$ what is the maximum absolute error in calculating $1/(x - y)$?

7

Storing and manipulating numbers within the machine

Objectives

After working through this chapter you should be able to:

❑ understand the way numbers are processed in the ALU
❑ see how the architecture of the machine dictates the storage and processing methods
❑ use different forms of coding for numbers
❑ understand how error is induced in the ALU.

7.1 Introduction

When first we learn to do pencil-and-paper arithmetic we quickly find that if we don't keep numbers in their proper columns our results soon become erratic. Sometimes we shall be lucky and keep the true values, on other occasions we shall be misled and make mistakes.

When we enter numbers into the machine we simply assume that they *will* always be kept in their correct relationships with one another. How do we know? We have already seen that error is often induced by the machine and propagated by our procedures. How do we know exactly what to expect?

7.1.1 We can assume that the machine will do its sums *reliably*, that if we tell it to subtract it will do so and will always do so in the same way. We can by no means take it for granted that it will do our computations *accurately*! Here is reason enough for computer staff at all levels to have a good grasp of what actually goes on within the ALU.

7.1.2 Some students of computing will also go on to learn to program in assembly language or machine code, the so-called 'low-level languages'. These, whilst being much more time consuming to write, are fast in execution and economical of storage space. To write such code we need a very intimate understanding of activities within the machine but the detail of the manipulations is outside the scope of this book.

7.2 Some preliminaries

7.2.1 First we need to link our ordinary arithmetic to what goes on in the machine. One of our problems when reading computer manuals is that they often speak of aspects or techniques of mathematics which we may never really have understood or which many of us have simply done mechanically, as we were taught.

Let us go back to some of these basic activities and relabel or relearn what is necessary.

7.2.2 We have seen in earlier chapters that arithmetic, which is all the machine can do with numbers, obeys certain laws. Among the most important of these, for our present purposes, are the laws of **inverses**. These have the effect that, for *any* calculation, we can achieve the same end by *performing the inverse operation on the inverse of the value.* Rather than *subtract* 3 from 5 we can *add* −3 to 5.

A second point to remember is that multiplication is really only **repeated addition** and division is no more than **repeated subtraction**. The simplest representation of 3 * 5 is 5 + 5 + 5 (*or* 3 + 3 + 3 + 3 + 3) but 3 * 5 is much shorter. In the same way we could write 12/3 as:

$$12 - 3; 9 - 3; 6 - 3; 3 - 3$$

and say that it took four subtractions to account for all the 12, or 'exhaust' it. A primitive pseudo-code model of division could be:

```
Accept M,N
Q ← 0
DO UNTIL M < N
  M ← M − N
  Q ← Q + 1
END DO
Display Q, 'remainder', M
END
```

7.2.3 Arithmetic also has a *language* of its own which has arisen from the fact that any calculation, no matter how simple, must have four elements:

(a) an indicator of what *operation* is involved
(b) a number to be *operated on*
(c) a second number showing the *'scale' of the operation so to speak*
(d) somewhere (in close proximity with (a) to (c)) to *display the answer.*

For example, 3 + 5 actually says something like, 'to a starting quantity, three, add (or count on) a further five, (or *increase* the 3 by 5)'. Unless we also say '... and tell me (or write down), what the total is' we shall be none the wiser.

The computer actually reflects this very accurately since if we merely enter, on the keyboard, 3 + 5, we shall never *know* if our instructions have been carried out! We must *also* enter something like 'display 3 + 5' in order to see the answer. Pocket calculators, on the other hand, are generally programmed to display the answer when we press the '=' button. We can deduce that this button performs a function similar to the 'display' command.

On paper we show the 'place for the answer' when we set out a sum like:

$$3+$$
$$\underline{5}$$

by drawing the line under the lower number. If we set out the sum in equation style, i.e. as 3 + 5 = we know that we are expected to put the answer immediately to the right of the '=' sign. We have learned similar 'drills' for setting out all our pencil-and-paper calculations, to the extent that when we meet a new kind of calculation one of our first questions is 'how do I set it out?'

7.3 Conventions

7.3.1 In mathematics we already have names for the *operations* and symbols to serve as shorthand for those names. We have 'multiply' to indicate repeated addition, for example, with *, × or · as the corresponding symbol. Much less familiar are the very ancient words for the three numeric parts of each of the four common kinds of calculation.

7.3.2 Since the operations are different the three words used in each case are also different:

❑ For addition the first number is the   **addend**
the 'second' and subsequent numbers is/are   **augend(s)**
the result of the computation is the   **sum** or **total.**

❑ For subtraction the corresponding parts are   **minuend**
and that which is to be subtracted is the   **subtrahend**
with the result of the operation, the   **difference.**

❑ Multiplication sees the rather clumsy   **multiplicand**
operated upon by the   **multiplier**
and producing, as a result, a   **product.**

For division, which we set out differently in pencil and paper calculations, we have

$$\text{divisor} \,) \, \overline{\text{dividend}}$$
$$\text{quotient}$$

Getting to know these is of no great importance in itself but they *do* allow us to use the one word rather than writing, each time, something like 'the number we are subtracting from'. They are of value while we are investigating our present topic but, happily, do not usually form part of everyday speech!

What we have learned in our schooldays of the 'setting out' of calculations, has included all these ideas but we have rarely learned them *as a set of fixed principles*. What usually happens is that we learn each 'drill' separately and rarely connect them into any coherent pattern. This often means that if we momentarily forget how to set out a particular problem we are defeated.

7.3.3 Even so we are much better at it than the computer which is actually incapable (with large and costly exceptions), of 'learning' anything. Much less can it learn and store different procedures for different kinds of sum except as program segments. Input devices do not even allow for such setting out. Indeed the machine will not even recognize the commas and/or spaces which *we* include in large numbers to make them easier to read. Where we might write 1,645,327 or 1 645 327, or some other notation which is familiar to us, the machine simply will not accept it.

We are left with a very rigid system which essentially recognises only the decimal point within a number and where all operations are conducted within the *same* framework, or 'setting out'.

7.3.4 If you have ever seen a mechanical calculator (see Figure 7.1) from the days before electronic machines took over you will have seen that they had three 'registers' in which the numbers involved in the calculations were placed or stored. These were very similar to rows of pigeonholes which **(a)** kept the numerals in their appropriate columns and **(b)** kept them in their proper 'rows'. One register held the number to be operated on, a second the augend, subtrahend, divisor or multiplier, whilst the third, generally called the **accumulator**, contained the result. There was also some device for setting the decimal point to indicate which of the columns contained the 'units'.

Since these were the only high-powered computational machines which mathematicians had been used to it was perhaps natural that they should have used the idea of registers in the computer when writing the earliest computer programming systems and control programs. They did make some changes since in the computer we commonly have at least six registers and they can be 'doubled up' for some operations. Nevertheless, in the ALU, it is

Figure 7.1 A mechanical calculator c.1960.

still essentially three sets of pigeonholes in the same relationship as the three 'rows' of a pencil-and-paper sum.

7.3.5 The computer, like the calculating machine above, does differ from our ordinary arithmetic, in truncating numbers for instance, since *its registers are of fixed length*. On paper we can keep adding columns to accommodate any size of number we choose.

7.4 Some less familiar arithmetic

7.4.1 Numbers have **complements**, just like sets and just as with sets they 'complete a whole'. For numbers in our ordinary base 10 sequence the complement is what is needed to make up the whole of the appropriate 'power of 10' which 'contains' the number concerned. For example 7' (the complement of 7) = 3 because 7 + 3 = 10; 437' = 563 since 437 + 563 = 1000. We describe these as the 'tens complements' of the numbers concerned.

Just as we can subtract a number by adding its inverse we can also do so by adding its complement and finally subtracting the '10s complement'.

7.4.2 For example: 5 – 3 = (5 + 7) – 10

$$800 - 437 = (800 + 563) - 1000$$

Check these, and then write several others, to make sure that they work and that you can operate with complements. Can you see that in the first example all that we are really doing is taking advantage of the fact that $-3 = +7 - 10$?

In our ordinary arithmetic this is little more than an interesting oddity although it helped speed up calculations with the hand calculator of days gone by. It does, however, have great importance in the operations within the computer.

7.5 Arithmetic in the computer

7.5.1 The computer *can only add* so that the performing of subtractions, within the machine, can only be done by adding complements.

7.5.2 The computer also only functions, at its heart, in binary numbers, so that all our base 10 numbers must be **encoded**, or converted to base 2, immediately after a numeric key is pressed on the keyboard. How this is done will vary from machine to machine and from application to application.

7.5.3 The essential is that all such codes reduce the numbers to the ones and zeros which obey the rules for binary: $0 + 0 = 0$; $0 + 1 = 1$; $1 + 0 = 1$; $1 + 1 = 10$; $1 + 11 = 100$ and $0 * 0 = 0$; $0 * 1 = 0$; $1 * 0 = 1$; $1 * 1 = 1$.

7.6 Binary coded decimal

7.6.1 For applications such as industrial process control, where values are constantly recorded and compared but only numerals are involved and little sophisticated computing is necessary, it is still quite common for **binary coded decimal** to be used, usually abbreviated to BCD.

7.6.2 In this coding system each numeral is given a four-bit binary code to correspond with the list of ordinary counting numbers from zero to nine. In this system each 'nibble', or half byte, of four **bits** is effectively kept separate (even if moved around the system in pairs as eight-bit bytes).

The codes are:

1_{10}	0001_2
2	0010
3	0011
4	0100
5	0101
6	0110
7	0111
8	1000
9	1001

leaving six 'spare' bits which do *not* represent numerals but which can be used

for other purposes if needed. For example, it is possible to use three of them
to represent the plus, minus and decimal point signs.

This is 'straight' BCD although more elaborate forms, including six and
eight-bit versions, have been developed for more complex tasks.

Arithmetic with BCD is quite straightforward since the digits are kept in
their nibbles and aligned in the registers just like ordinary numbers.

7.6.3 For example, 473 + 426 would be

$$
\begin{array}{ll}
0100\ 0111\ 0011 & \text{in the addend register,} \\
0100\ 0010\ 0110 & \text{in the augend register} \\
\hline
1000\ 1001\ 1001 & \text{in the accumulator}
\end{array}
$$

If we translate the result back into denary we see that $1000_2 = 8_{10}$; $1001_2 = 9_{10}$ and $1001_2 = 9_{10}$ or 899. This example does not involve any 'carries', i.e. no total, for any individual column, >1001, or 9_{10}.

7.6.4 Let us see what happens when we *do* have carries.

597 + 458 becomes

$$
\begin{array}{ll}
0101\ 1001\ 0111 & \text{addend} \\
0100\ 0101\ 1000 & \text{augend} \\
\hline
\qquad\qquad\ \ 1111 & \text{partial sum}
\end{array}
$$

7.6.5 (Notice how the partial sums, intermediate results, build up in the
accumulator.) Notice too that the sum of the units column, the rightmost
column of nibbles, = 1111 which *is not* a 'legal' code in BCD where the largest
possible numeral is 1001.

We must then *restore* the result to the legal maximum by *subtracting* the 1001
from 1111, leaving 0110 which we must enter and 'carry' 1 to the next column,
to give:

$$
\begin{array}{ll}
0101\quad 1001\quad 0111 & \text{addend} \\
0100\quad 0101\quad 1000 & \text{augend} \\
\hline
\qquad\qquad\quad\ \ 0110 & \text{partial sum} \\
\hline
\qquad 0001 & \text{carries}
\end{array}
$$

which in turn gives the next column sum as 1001 + 0101 + 0110 = 0101 with

0001 as the carry so that, at the end of this stage we see:

0101	1001	0111	addend
0100	0101	1000	augend

	0101	0101	partial sum

0001	0001	carries

When we add the final column we have $0101 + 0100 + 1011 = 1\,0000$ to give, as a final result

0101	1001	0111	addend
0100	0101	1000	augend

0001	0000	0101	0101	accumulated sum

0001	0001	0001	carries

which we can translate back into 1055_{10}.

7.6.6 The leading zeros on the 'carries' are actually redundant. I have included them at this stage only to make clear that it is the *binary* one which is carried and that it is 0001 in its four-bit, BCD form.

As always you should practise such BCD additions until you are confident with them.

7.7 More sophisticated codes

7.7.1 If we are dealing *only* with numbers this is a very simple procedure and could be handled by a fairly elementary processor. It is too slow and takes up too much space for more general use.

7.7.2 Modern information processing demands much more of our systems than merely dealing with numbers and the American Standard Code for Information Interchange (ASCII), developed from the very old Baudot code, or better still the eight-bit extended ASCII, provides for processing far more than simple numerals.

7.7.3 IBM, which has always preferred to develop its own systems rather than adopt existing standards, took the extended BCD concept from four- to six- to eight-bit and eventually into its extended binary-coded decimal interchange code, EBCDIC, still largely restricted to IBM machines. Both ASCII and EBCDIC were developed for handling verbal as well as numeric data, but most of the complexities of handling text need not concern us greatly in this book.

7.8 Capacity and register size

7.8.1 The architecture of most modern computers, the whole of their internal communications, processing, control and storage, is based upon the 'word-size' of the system. Very small computers have eight-bit processors, medium-power machines have 16-bit architecture and high-power machines have 32-bit systems. (Note that in this case the power of the machine is related to speed and capacity not its electrical consumption.)

7.8.2 From our point of view this means that each of the registers holds 8, 16 or 32, bits respectively. For commercial and industrial purposes eight-bit machines are rather slow and limited in memory size. In principle a set of 32-bit registers functions in exactly the same way as do 16-bit registers. For the sake of simplicity I propose to take a brief look at the key difference resulting from register size.

7.8.3 Within an eight-bit machine the largest integer we could store in this way would be 01111111 and the smallest 10000000, since this is the smallest possible value in twos complement form. The largest possible value then is +255 and the smallest –256; not much use for serious arithmetic!

7.8.4 The remedy is to use eight-bit registers in 'pairs' for all arithmetic operations, giving us 16 bits for addend, augend and accumulator registers and largest and smallest values of 0111111111111111, or +32767, and 1000000000000000, –32768, respectively.

By the same process a 16-bit machine, with paired registers, can hold +2,147,483,647 and –2,147,483,648 in integer form, which allows us to deal with quite large numbers.

7.9 Induced error

7.9.1 When using the machine there will, we must remind ourselves, be frequent induced errors from the decimal/bicimal fraction conversions, known as **conversion error**, which we looked at in Chapter 5.

7.10 Conclusions

The nature of the machine makes many of our ordinary mathematical techniques inapplicable but, subject to the laws, we can use others which precisely suit the operation of the ALU.

New techniques need new understandings and experience if they are to become truly familiar to us. They also introduce new errors and limitations.

Exercises

(1) Find the tens complement of:
  (a) 563
  (b) 999
  (c) 101
  (d) 23.
(2) Using tens complements calculate:
  (a) 499 − 367
  (b) 499 − 637
  (c) 2495 − 792
  (d) 1207 − 649.5.
(3) Code in BCD:
  (a) 495
  (b) 222
  (c) 632
  (d) 410.
(4) Calculate in BCD:
  (a) 493 + 275
  (b) 493 − 275
  (c) 47 + 39 + 63
  (d) 547 + 659 + 237.
(5) Convert to base 10 the following BCD numbers:
  (a) 0111 0011 0001 1001
  (b) 1010 0001 0010
  (c) 0100 1010 0110 0110
  (d) 0111 0101 1011 1100.
(6) Express the following in eight-bit, twos complement form:
  (a) 59
  (b) 44
  (c) 73
  (d) 37.
(7) What is:
  (a) (i) the largest
       (ii) the smallest number
       which may be held in a 16-bit register?
  (b) (i) the largest
       (ii) the smallest number
       which may be held in a 32-bit register?
(8) What is the practical importance of the difference in the answers to question **(7)**?

8

Graphs, functions and inequalities

Objectives

After working through this chapter you should be able to:

❑ understand the need for graphical skills in computing
❑ produce graphs to a high standard
❑ generate graphs using a spreadsheet
❑ understand the significance of scales
❑ relate functions to their appropriate types of graph
❑ model continuous, linear, curved and scatter graphs
❑ write appropriate computer program segments
❑ use simultaneous equations and inequalities in graphs
❑ assess error in graphs and graphical techniques.

8.1 A new need

8.1.1 The widespread use high-resolution visual display units (VDUs), dot-matrix, laser and ink-jet printers, powerful software, fax machines and high-speed, high-fidelity data transmission systems have, in recent years, transformed the world of information processing. Today visual models, as distinct from 'traditional' numeric or verbal outputs, have become common-place and a *major working tool for users at all levels*.

8.1.2 The old saying that, 'a picture is worth ten thousand words...' is especially true of mathematics. Most people find difficulty in reading and interpreting large quantities of *numerical* information, even if it is arranged in the form of tables. Traditional computer print-outs have been especially difficult to read and people can see the implications of that information much more clearly if it is presented as a picture, which is all that the word 'graph' means.

At a more sophisticated level we may construct graphs of relationships which allow us to derive considerably more information than we actually *put into* the graph whilst constructing it, or to solve quite subtle problems. At this level the graph can provide us with an immensely powerful tool for analysis and prediction and the computer is well suited to generating, displaying, storing and transmitting such graphs.

8.2 Pictograms

8.2.1 Some of the world's earliest works of art, sophisticated paintings found in caves all over the world, may simply be graphs; records of the game found in the hunt for food or of the animals domesticated and used by the peoples who left us their pictures. Even today we still draw small images of what we want to record and when we group such images for the purpose of displaying or comparing situations we call the resulting graph a **pictogram** (see Figure 8.1).

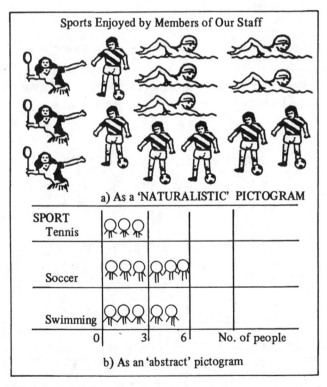

Figure 8.1 Two forms of pictogram.

8.2.2 The images may be quite realistic or severely abstract but they confront us immediately with one of the basic problems of all graphs, indeed of all pictures. Although people often say, quite wrongly, 'figures cannot lie' we seldom add, '... but pictures can and frequently do!'.

8.3 Distortion

8.3.1 All pictures are themselves abstract models of what they represent and the person producing the picture makes decisions, consciously or unconsciously, about what to include, simplify, distort, omit or scale. Every artist does this whatever medium is being used and even in photographs the finest camera in the world still introduces its own misrepresentations into the images it creates.

8.3.2 If the person creating the image is a politician or an advertiser the distortions may be intentional. When we look at any graph we should ask ourselves whether the originator is simply modelling facts or has some intention of trying to persuade us to believe or buy something! This can be seen in Figures 8.2 and 8.3.

8.4 Rules for graphs

8.4.1 In order to present graphs which are both useful and worthy of respect we must follow certain mathematical rules, some of which are listed here. We will add others as they become necessary.

Figure 8.2   Truth or propaganda?

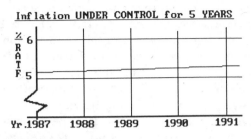

Figure 8.3   Government or opposition?

(1) The type of graph chosen must be appropriate to the material being modelled.
(2) We must give the graph a title which shows clearly what it is displaying.
(3) We should add any subtitles needed to clarify its purpose.
(4) We must indicate clearly whether it is a one-, two- or three-dimensional representation.
(5) We must define the scales of measurement used on each of the axes.
(6) All scales must be clearly labelled and must be true to the range of values represented.
(7) All calculated or input values must be clearly indicated and must be distinguished from *derived or inferred* values.
(8) Unless otherwise stated **time** or the **known** values (often referred to as the 'x' values), will be displayed along the *horizontal axis* with, for example, **distance** or the **unknown**, ('y') values allocated to the *vertical axis*.

   8.4.2 Let us look in a little more detail at the different *types* of graph and what each is suitable for. Computer software packages rarely provide for true pictograms but, in any example similar to our sports graph in Figure 8.1 we can represent each person, more simply, by a 'box' on a **block graph**, as shown in Figure 8.4.

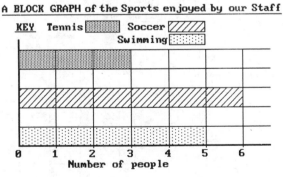

Figure 8.4   A block graph of the information in Fig. 8.1.

Note that in this case each figure is represented by a box along the same horizontal axis as was used in the original graph in Figure 8.1. There are a number of important things to note about this simple graph. It is, although drawn in two dimensions, a *one-dimensional* situation. Only the *number* of people in each category is significant and the categories must be indicated by either separate labels or by a 'key'. For ease of comparison the blocks all start from the same 'baseline' or **datum**, using the vertical axis for this purpose although the true **origin** of the graph, since it is one-dimensional, is the left-hand end of the horizontal axis. This is done *by extending the zero datum vertically to serve as a common base.*

8.4.3 This is *not* at all the same as having a true vertical scale. The categories, tennis, soccer, swimming, which are *data labels*, rather than data, are also separate from each other, in mathematical language they are **discrete**. We must, where this is the case, keep the piles of blocks separate from each other.

8.4.4 *Only where we are using a truly continuous scale may we place the blocks in contact with each other.* Where this is the case the resulting graph is a **histogram**, of which we shall see a good deal more later, which is not at all the same as a **block graph**. In order to make our graph look more familiar to us we may, instead of keeping to the original layout, make 'the floor' our datum line by plotting the graph with the *number of people* on the vertical axis. This is the more general arrangement for a block graph simply because gravity obliges us, in reality, to pile blocks in vertical, rather than horizontal, columns. Block graphs are essentially easy to draw and very clear. Have a look at Figure 8.5. They are a little misleading since they suggest that it is the *area* of the block which is important. It is not. Only one dimension actually represents anything

Figure 8.5   A block graph 'right way up'.

significant (and we should note that in our example this one true scale also gives *discrete* values, in this case integers, since we cannot have less than one whole person).

8.4.5 At the lowest level of abstraction our original individual people may be represented by a simple line *whose length indicates the single dimension, number,* with which we are concerned. This type of graph also may be drawn with the lines either *horizontal,* usually called a **bar chart** (or graph), or *vertical,* when it is often referred to as a **column graph** (see Figure 8.6).

Figure 8.6   The same information simplified yet again.

8.4.6 These three types, the **pictogram, block graph** and **bar chart** are simply *display* graphs; they model a fairly simple situation, are built up by plotting *all* the relevant data and are *interpretive only.* We cannot infer from them anything which is not explicitly stated on the graph. Note, however, that our rules **(2)** to **(6)**, above, must be obeyed in constructing them if we are to give the viewer a true picture of what is displayed.

8.5 Spreadsheet graphs

8.5.1 *We* must understand and get used to these rules because any spreadsheet system which has graphical capability will call on *us* to identify and input all these pieces of information before it will prepare the relevant graph. It may also not obey all the rules precisely and we may need to alert the user to any such peculiarities.

8.6 Proportional charts

8.6.1 Another type of graph which is commonly found in spreadsheet systems is the **proportional chart**. In this case the graph shows the whole of

COSTS(%)	1989	1990
Fuel	5	10
Materials	20	25
Overheads	20	15
Wages	40	35
Advertising	5	3
Salaries	10	12

Figure 8.7   A table of costs for two years.

some entity, say *costs*, as in Figure 8.7, and subdivides this into the proportions contributed by various subdivisions. When working with proportions
it is most convenient to express the parts as *percentages* as has been done here.
This permits direct comparisons, year-by-year, or between similar companies,
or divisions of a large organization, in a way which is not nearly so easy if the
various parts are expressed as common, or vulgar, fractions. If in one division
of a company wages are 17 out of 23 $M or 17/23 of total costs, whilst in
another they are 19 out of 25 $M, (19/25), total which has the proportionally
higher wage-bill? If we convert the fractions to 74% and 76%, respectively the
comparison is easy.

8.6.2 A simple way to graph such a table is to create a **proportional bar chart**
where the data are displayed two-dimensionally (for ease of labelling, shading
or colouring the different proportions), but only the length (or height), is
truly significant as the dimension which represents each of the individual
percentages of the whole.

8.6.3 As we can see in Figure 8.8, the information is displayed clearly
enough but it is not very easy to read from the graph the full *detail* of the
changes which have taken place. We *can* see, for example, that the first three
items have risen, in total, to 50% of overall costs and that proportionally less
has been spent elsewhere. But it is difficult to assess from the graph exactly
what has happened to the last three items of costs. A more detailed scale might
help.

Figure 8.8   A proportional bar chart of our data.

8.6.4 Another very popular form of the proportional graph is the **pie chart**. In this the different proportions are allocated correspondingly sized 'slices' of the whole pie. When we make such a chart we need to remember that the angle at the centre of a circle is 360° so that, if we are working with percentages, each 1% = 3.6° at the centre. Even this gives rise to some very awkward angles if we are trying to measure them by protractor, but vulgar fraction proportions can be very much more awkward. How many degrees at the centre would 17/23 represent?

8.6.5 The result is that measurements on pie charts are always, in practice, likely to be approximations.

8.6.6 In Figure 8.9 we see that, once again, it is difficult to read the *exact* significance of the different sizes of sectors even if we do have a protractor handy when we are attempting to interpret the graph and this is unavoidable.

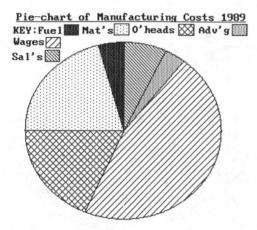

Figure 8.9 Our data in a pie chart.

8.6.7 Such graphs give a reasonable impression of the data from which they were constructed, but are more suitable for display purposes rather than for any subtle interpretation. We can improve matters somewhat by *annotating* the graph, as shown in Figure 8.10 on the next page, by writing in each sector the proportion which it represents. In practice almost all spreadsheet systems automatically do this to make the graphs more legible to the user.

8.6.8 A final form of this type of graph is what is known as an **exploded pie-chart** where, in order to emphasize the significance of one or more elements of the graph we show the 'slice' or slices concerned as partly

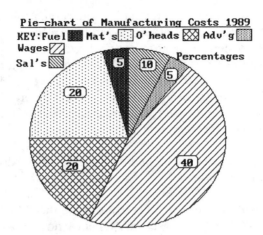

Figure 8.10   An annotated pie chart.

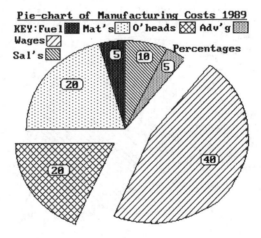

Figure 8.11   An 'exploded' pie chart.

withdrawn from the whole pie. This can be seen in Figure 8.11. These are slightly more difficult to draw by hand but many spreadsheet systems offer the facility.

8.6.9 It is worth noting that some writers, clearly believing that there can be nothing simple about mathematics, insist on calling these 'pi' charts, using the verbal form of the Greek symbol π, presumably because this has something to do with circles! This has nothing at all to do with the nature of these graphs.

8.7 Scales

8.7.1 Before passing to other types of graph we need to look at the matter of scales (item **(5)** on the list in 8.4.1). In order to do this we must also examine more closely the nature of measurement, which may be on a **nominal, ordinal, interval** or **ratio** scale.

8.7.2 For a nominal scale we are using numbers simply as labels under which to *classify* items, in a very common but rather deceptive way since it has nothing to do with ordinary arithmetic. We could say, for example, that in our survey of sports, all tennis players should be put into group 1, all soccer players into group 2 and all swimmers into group 3 and we could then number the blocks or bars of a graph accordingly. Such numbers do not indicate any measurement and certainly do not imply that group 1 is superior to, or even comes in any sequence *before*, group 2, so we must be extremely careful always to treat them as just the *names* of classes and *not* as true numbers.

8.7.3 Such measures are often useful, especially in some kinds of statistical tests, but must always be treated as **non-parametric**.

8.7.4 If we take a different approach and ask 100 people to classify different sports *in order of preference* and the verdict emerged, tennis 1, soccer 2, swimming 3 it is clear that, unlike in the **nominal** scale, the sequence *is* important since it has **ordered** the classes. This is the **ordinal** scale of measurement which we frequently use in order to **rank** or **order** items. Again it is non-parametric and it does not indicate by *how much*, if such a thing even *could* be measured, people prefer tennis to soccer. It does not **quantify** our preferences.

8.7.5 When we make a **scatter graph** in statistics we may plot, for example, students' **rankings** for one subject against their rankings for another subject, using an *ordinal scale* on each axis.

8.7.6 When we prepare graphs we must be extremely careful that our presentation does not *imply* that either nominal or ordinal measures alone may be used to **infer** anything more than the bare facts displayed on the graph.

8.7.7 We move into a different order of measurement if we ask the same people to *rate*, on a scale from 1 to 10, their *enjoyment* of those same three sports. It is then implied that each **interval** on the scale has the same magnitude and that someone rating tennis as 10 and soccer as two would consider the former to be five times as pleasurable as the latter.

8.7.8 Rather similarly we would be entitled to assume that someone with an intelligence quotient, (IQ), of 110 was as far *above* the 'norm' of 100 as is someone with an IQ of 90 *below* that same norm. These are two applications of an **interval scale** of measurement, where there is exactly the same *interval* or range of value between any two marked points.

8.7.9 The importance difference is that the example in 8.7.7 is **subjective**, or is a matter of feeling or opinion, whilst in 8.7.8 the values, which are based on actual *measurement*, are **objective**. We must always recognize **and declare on our graphs**, by means of a subtitle perhaps, if the information being displayed is *subjective*.

8.7.10 Some interval scales, the IQ scale for instance, take a mid-point or 'norm' and spread out on either side of that. There is no 'zero IQ', only a number which is more or less than the norm of 100. We find many such scales in statistical tests and graphs and shall see more of them in that context later.

8.7.11 The most refined form, or the **highest order** of measurement is that on a **true ratio scale**. For all of these we deal in **objective values** on a scale which has a **true zero**. Virtually all measurements made with instruments are of this nature and many of them originate at zero and continue to infinity. They may also extend to the negative as well as the positive numbers.

8.7.12 The final point about all such scales is that they may be **discrete** or **continuous**. When we are counting the number of people at a football match, for example, we are using a **ratio** scale, since there can be a true zero, but each person is a separate entity and there are no fractions. We use a **discrete ratio scale**. If, on the other hand, we are plotting a graph of a sick patient's temperature the mercury in the thermometer can vary infinitely between say 32 and 33°C and the variations are on a **continuous ratio scale**.
This is one of the most important and least understood aspects of measurement and consequently of both the presentation and interpretation of graphs. *Many spreadsheet and other graph-drawing packages are quite unable to distinguish between discrete and continuous data. If this is the case we* should use a subtitle to draw attention to the true nature of our graphs.

8.7.13 Let us look at a specimen graph in order to clarify the essential differences between a discrete and a continuous scale. If we take the sales chart in Figure 8.12 we see that, represented as a bar graph, we are looking at a *discrete* scale on the horizontal axis. The vertical axis is a true continuous ratio scale since any amount may be represented.

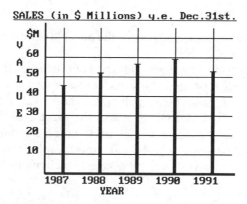

Figure 8.12 Sales as a bar chart.

8.7.14 When we display the same data as a block graph, however, since there seems to be a continuous time-scale from 1987 to 1981, it might seem that we could place our blocks side-by-side as in Figure 8.13.

8.7.15 This is not strictly true since the scale is discrete and there is actually a gap of 364 days between 31 December 1987 and 31 December 1988 – and likewise throughout the horizontal scale. The totals for sales are not recorded day-by-day in our data (although they may have been within the accounts department of the company concerned). All that we have is the **total of sales**, accumulated throughout the year, without any knowledge of their details. This also means, as we saw earlier, that it is only the *heights* of the blocks which is significant.

8.7.16 Even if we do close the gaps between the gaps, as in Figure 8.13, this *is not a* **histogram**, it is still only a block graph.

Figure 8.13 The same data as a block graph.

8.8 Points and trends

8.8.1 There is another consequence of the discrete scale which is important to us in this example.

We could equally well, as in Figure 8.14, plot our data on a graph by marking *the exact points* which represent the values of sales for each individual year. We usually mark each point by a small cross whose exact intersection is the value concerned, or by a dot which marks the exact position but with our attention drawn to it by placing a small circle round the point. (Note that it is not uncommon for spreadsheets and other graph-drawing software to identify the point by drawing a small square or a circular blob rather than a precise point.)

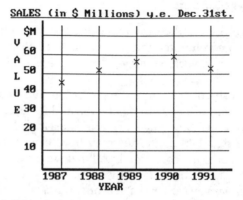

Figure 8.14   A 'point' graph.

8.8.2 Having plotted our points we must not, however, then 'connect' them with an *unbroken* line. This is because there is no evidence whatever that sales rose *either steadily or continuously* between 31 December 1988 and 31 December 1989. They will in all probability have fluctuated between any two recorded dates, perhaps considerably.

8.8.3 An unbroken line defines the *relationship* as **continuous** and this has a special meaning as we shall see more clearly when we look at the graphs of functions presently.

8.8.4 All that we may say, and we must use clearly *discontinuous* lines on our graph in order to say it, as in Figure 8.15, is that *there was a general trend for sales to rise* between those two dates, taking the year as a whole. This is another of the strict rules which is often broken by spreadsheet systems which tend to treat *all* line graphs as continuous.

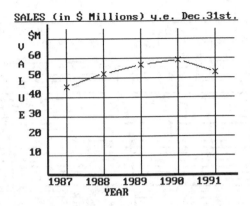

Figure 8.15 'Trend' lines for discrete data.

8.8.5 Such 'trend-line' graphs are actually of great value in management and we may, at a more sophisticated level, analyse and continue the trend beyond the known data.

8.9 Continuous functions

8.9.1 We may only use a continuous line on a graph when we have clear evidence of the *continuity* of the function but in that case we *must* use a continuous line, which truly follows the function. Providing these rules are obeyed we may then **interpolate** (read values *between* plotted points), and **extrapolate**, or read values *beyond* the plotted points.

8.9.2 The example shown in Figure 8.16, a chart from a recording barometer, is of truly continuous ratio scale graph. Here the pen which draws the line on the graph, indicated by the arrow, is continually moving in the y axis because the pen is mounted on a swinging arm which rises and falls, in real time, as the atmospheric pressure changes. Along the horizontal, x, axis the paper, wound on to a drum, is moved by a clock mechanism, also in direct relationship to real time.

In this case the variation in height along the y axis is a naturally occurring phenomenon and cannot be simply predicted. This is no simple two-variable relationship but the very complex association of many variables in the world weather system.

8.9.3 Such continuous graphs are relatively common in computer-controlled industrial processes where the machine is continuously reading two or more parameters and recording them by plotter or by a continuous line of overlapping dots.

Figure 8.16   A truly continuous graph – in both dimensions.

8.10 Graphs of functions

8.10.1 Happily, at this level we are not called upon to deal with such complicated models but there *are* three classes of simpler two-variable functions on continuous ratio scales which we can and should be able to model.

8.10.2 They are:

(1) $y = mx + c$ (first order)
(2) $y = m(x^2) + c$ (second order)
(3) $y = m(x^3) + c$ (third order)

and variations on, or combinations between any of the three.

8.10.3 The three types of function, the **linear**, **quadratic** and **cubic**, respectively, have much in common since each of the values of x may be multiplied by the value *m*, which may be positive, negative or unity. Each may also have added a constant value, *c*, which may again be positive or negative but may also be zero. You will, I am sure, remember that in the laws of **identity** one and zero are the identity elements for multiplication and addition respectively; they leave the value of the *x* unchanged.

8.10.4 In:

$$y = 2x \qquad\qquad \text{the } m = 2 \text{ and the } c = 0$$
$$y = x - 5 \qquad\qquad \text{the } m = 1 \text{ and the } c = -5$$
$$y = 2x^2 - 7 \qquad\qquad \text{the } m = 2 \text{ and the } c = -7$$

and so on. The *m* indicates the steepness of the curve on the graph or the *ratio* or gradient (the rate of change) of the slope, often expressed as a **ratio**. For example that ratio will be 4:1 if the *y* value increases at *four times the rate* of the change in *x* value. In that case *m* = 4.

8.10.5 The *c* indicates the height, above or below zero, at which the curve 'cuts' the *y* axis; this is known as the *intercept* or *displacement*.

8.10.6 We may also have various combinations of elements as in

$$y = 3x^2 + 2x - 7$$

which do not add any great complication to the model but may look a little more forbidding.

8.10.7 One such function which looks a little complicated but produces a slightly surprising result when plotted is of the type $x^2 + y^2 = 4$. Try this one later and see what happens.

8.10.8 The three types of function show characteristic curves on the graph fields (which can be seen in Figure 8.17) and their actual steepness and positioning will vary according to the actual parameters of *m* and *c*.

FIRST order   SECOND order   THIRD order

Figure 8.17   The three characteristic graph curves.

8.11 Some further rules

8.11.1 We can, once again, draw up some general rules for the presentation of the graphs.

(1) Unless we cannot have negative entities in the universe of discourse concerned the scales on both axes should embody at least some part of the negative values in their scales.

(2) Unless the set of values on the scale concerned is *known to be finite* the scales should be shown, in each axis, to extend to infinity and be continued right to the edge of the graph paper.

(3) *The scale chosen for the graph should be the largest which will cover the range of values stipulated on the size of sheet available* so that we may have the benefit of the largest plot which can be fitted on to the paper. (If, for example, we are told that the x values should extend from −3 to +3 the whole width of the x axis should be used for, say, 7 units so that we may have as accurate a graph as possible.)

(4) The marking of points and the drawing of lines should be as fine as possible to minimize induced error. The ideal combination is the largest possible graph and the finest possible lines, including the printed scale.

(5) For a linear function three points should always be plotted as a check on the arithmetical accuracy of our calculations. If the three points do *not* fall accurately on a straight line one of them has been wrongly calculated.

(6) For all second- and third-order functions the line of the graph, no matter how few points have been plotted, is a *continuous smooth curve. The points must on no account be joined by straight lines.* The more points we plot the easier it is to smooth the curve, especially with spreadsheet systems where lines between plotted points almost always break this rule by being straight! (This is due to a limitation in the software which we cannot ourselves correct.)

(7) We should label the curve on the graph with the function which it represents. This becomes especially important when we have two or more functions plotted simultaneously on the same graph field.

(8) If and only if all these conditions have been complied with can we use the graph for reliable interpolation and extrapolation.

8.11.2 For the actual plotting of points it has long been customary, and still is for *manually* drawn graphs, to set out the calculation of the points for plotting a function $y = 2x^2 + 2x - 3$ such as is shown in Table 8.1.

8.11.3 This method is still very reliable and, as with the $y = mx + c$ graph, if any one of the points does *not* fit to our carefully drawn *smooth curve* it is likely to be the arithmetic which is at fault.

Table 8.1

If	$x =$	−3	−2	−1	0	1	2	3
then	$2x^2 =$	18	8	2	0	2	8	18
and	$2x =$	−6	−4	−2	0	2	4	6
and	$-3 =$	−3	−3	−3	−3	−3	−3	−3
hence	$y =$	9	1	−3	−3	1	9	21

8.12 Graphs by computer

8.12.1 Such an elaborate table is not needed for the computer nor is it desirable. Remember that the great advantage of the machine is that it will, without becoming bored or impatient, do large numbers of repeated calculations in very little time. This has many advantages since we can calculate many points and derive a much smoother curve from a machine even though it resolutely tries to draw *straight lines* wherever possible.

For the example above, where we wish to plot our curve for values of $-3 \leq x \leq +3$, we could as easily take intermediate values at intervals of 0.01 as of 1. The 600 separate calculations would take little time in the machine.

```
X,Y REAL
DIM X(601); Y(600); j = 1; X₁ = −3;
DO UNTIL Xⱼ = 3.01
    Yⱼ ← 2*(Xⱼ ^ 2) + (2 * Xⱼ) − 3
    j ← j + 1
    Xⱼ ← Xⱼ₋₁ + 0.01
ENDDO
DO PLOT
END
```

is just one sketch of a pseudo-code routine which would calculate all the values in readiness for calling a 'plot' routine to print the curve either on screen or on paper.

8.13 The area under a curve

8.13.1 A long-standing problem with such 'curved' functions has been to find the area under the curve between any given pair of values and the traditional calculus gave us, very laboriously, approximations of that area. By a small modification of the routine, suggested above, for plotting the points we can dispense with such manual labours today.

8.13.2 Between any two successive values of x the area of the narrow vertical strip beneath the line of the graph is effectively a rectangle whose width is the constant 'step' size (i.e. the difference between the two x values) and whose height is the average of the two corresponding calculated y values.

We can append this additonal calculation, of the area of each strip, to the existing routine and on each pass of the loop add each new area to a variable, say A, as a **cumulative** total. Try such a routine for yourself and then code and run it on a computer.

The computer, with its high-speed iterative methods has in such ways rendered the calculus virtually obsolete for most practical purposes.

8.14. Interpolation and extrapolation

8.14.1 From any of our plotted continuous functions on a true ratio scale we can derive the values represented by any other point on the line. For any given value of x we raise a vertical line to the point where it intercepts the curve of the function and by projecting a horizontal line to the y axis we may read off the corresponding y value.

8.14.2 By the inverse process we can find, from any known y value, the x value from which it is derived. Since this is a quite common requirement it is often well worth adding such an 'enquiry' routine to the program to generate the curve of the function. Whether or not we wish to show such interpolations on the VDU screen will influence the way in which we set up such routines. You should experiment with such procedures.

8.15 Simultaneous functions

8.15.1 The next important development for us is when, for a given situation, two or more functions are simultaneously true.

8.15.2 If, for example, we have two equations of the $y = mx + c$, first-order type in which the m is different in each case, each will have a straight line on the graph and, if the functions are infinite those lines must intersect *somewhere* on our graph field so long as we have chosen the correct scale to reveal that intersection. You can see this in Figure 8.18.

8.15.3 The unique point at which the lines intersect will be the one value of the functions which is true for both. There are one or two other points to note about *this* particular figure for which the graph was generated using a spreadsheet system. See also Figure 8.19.

Figure 8.18 The intersection of $y = 2x + 7$ and $y = 1/4x - 10$.

Figure 8.19 The graph of $y = 2x + 2$ and $y = x^2 - 5$.

8.15.4 The first point is that the scales are shown at the left and bottom edges of the plot and *not*, as we would when constructing the graph *manually*, passing through zero and subdividing the graph field into the four quadrants.

The second is that the type of graph and its actual specification had to be input through the quite elaborate system of menus, some ten of them each presenting a considerable number of choices among which *we* have to make all the necessary decisions as to precisely how the graph is to be set out.

It is essential when we use such systems that we should have a real understanding of the underlying principles of graphical modelling.

8.15.5 If one is a linear and the other a second (or third) order, function there may be more than one point of intersection. In Figure 8.19, for example, where we have the plot of *both* $y = 2x + 2$ *and* $y = x^2 - 5$, we see that there are two points of intersection and in that case *each of those points* will be common to the two functions and the common values of x, y or both may be read off the graph. See Figure 8.20 for some further examples.

Figure 8.20   Points where two functions are true at the same time.

8.15.6 Once again you should note that a considerable number of points has been plotted, in order to make the curve (drawn by the spreadsheet system as a sequence of short straight lines), as smooth as possible. It follows, from the relative inflexibility of the spreadsheet software, that the same points are used to plot the *straight line* function, but only three are truly necessary.

8.15.7 Can you also see that any curve may, for analytical purposes, be taken as an infinite number of infinitely short straight lines?

8.16 Error in graphs

8.16.1 Any graph is subject to limitations on interpretation due to both inherent and induced error. There is inherent error in the numbers used, just as in any other application, *but there is also error induced by our technique.*

8.16.2 In all cases we shall find that large-scale plotting is needed to minimize induced error since both the thickness of the lines and the angle at which the lines cross may induce prohibitively large error if the graph is on too small a scale. Figure 8.21 shows the very small area of uncertainty when very fine lines cross at right angles, or close to it. It is so small that in the diagram it has been 'magnified' to three times its true area to make it clear. On the right of the same diagram we see the very large area of uncertainty, the error bound, which is induced if badly drawn, *thick*, lines cross at a 'shallow' angle.

Figure 8.21 Error induced.

8.16.3 The larger the scale the smaller the error induced by the thickness of the lines and the closer the 'crossing' of the lines is to a right angle the smaller the induced error at the intersection. Even the lines created by the printer of the graph paper induce error and it pays, firstly to use the very best quality of graph paper for the really important graphs and secondly, not to claim too high a level of accuracy if the printing of the graph paper is poor.

8.16.4 In practical terms, in computer-generated graphs, we need to be able to take account of the inherent dot size of the printer, the accuracy of the movement of the paper when printing and the number of dots per centimetre or inch which can be printed, both vertically and horizontally.

These things come under the heading of 'resolution' when we talk of print quality and accuracy and it is well worth becoming familiar with the capability and the limitations of the printers and the screens we are using.

8.17 Inequalities

8.17.1 The last aspect of graphs, for the present, is the displaying and solution of problems concerning simultaneous *inequalities*.

8.17.2 We saw, much earlier in the book, that any line on a graph embodies the set of points which satisfies the equation, or the *equivalence function*. So the graph of $y = 2x + 2$ will embody, in the line we draw, the infinite set of points for which that equivalence holds good.

8.17.3 Every such continuous ratio scale function has its own infinite set of such points. By defining such a set we also inevitably define the *in*equalities which are also created. They are:

(a) $y > 2x + 2$
(b) $y < 2x + 2$
(c) $y \le 2x + 2$
(d) $y \ge 2x + 2$

of which the last two *include* the line of the function itself whilst the first two are the regions, or domains, *above* (but *excluding* the line), and *below* (but excluding), the line, respectively.

8.17.4 There are, in real life, many problems where it is simultaneously true that two or more functions of the '*not more than*' type govern a situation, often by setting or defining the **limits** or **boundaries** of what is possible or **feasible**.

8.17.5 In Figure 8.22, we see three linear functions which in this example define limits which *may not be exceeded*.

Figure 8.22   The boundary of a feasible region.

For function 1 there is a fixed maximum value of $y \leq 1440$ whilst each of the other is of the 'inverse ratio between x and y' type. The precise nature of the functions need not concern us in this example of such a model but what *does* concern us is that the area under and including each of the three lines represents the 'not more than' region in each case so that *only the region which is below and including (sections of) all three lines together satisfies all three constraints simultaneously.*

8.17.6 The three function lines together define a polygon within which any possible feasible solution must lie. We call this, in practice, the **feasibility region** (or **polygon**). If, using ordered pairs for the points which define the polygon, we start at (0,0) and follow the sides of the polygon clockwise the subsequent points are approximately: (0,1440); (135,1440); (300,1000); (540,0) and back to (0,0). Check these and be sure that you can clearly define the region.

8.17.7 Many management problems can be modelled in this way on a spreadsheet system and explored on a 'what if?' basis by varying any one of the functions at a time. Such planning or forecasting problems are often constrained by several 'not more than' functions, manufacturing capacity, for instance. The great advantage of using a computer spreadsheet graphical model is the ease with which we can change one or more of the functions and quickly see the effects. We can then print out the graphical models so that managers can *see* the area of manoeuvre which is open to them. (Good experimental practice demands that only one variable should be changed in any experiment.)

8.18. Conclusion

Graphs are a vivid means of modelling real situations, give us a powerful tool for analysing situations and allow us to interpolate or extrapolate solutions to many different kinds of problem. They fit well with the computer display on the VDU and with modern high-resolution print-out.

Exercises

(1) Draw appropriate graphs to display the following:
  (a) A child's height in centimetres:

Age	2	3	4	5	6	7	8	9	10	11	12	13	14
cm	87	98	107	112	117	123	127	131	135	140	145	149	154

  (b) A company's sales, over a 16-week period:

Week	10	11	12	13	14	15	16	17	18	19	20	21	22	23	24	25
Sales (000s)	3.2	2.3	4	4.3	3	4.8	4.4	4.4	3.6	5.2	3	4	2.9	4.6	4.7	5

(2) (a) Comment on the graphs in **(1)**.
  (b) Would it be proper or valuable to interpolate on either graph?
(3) (a) Over a three-month period the earnings in a garage were $37,500 from petrol sales, $21,100 from repairs, $1200 from the car hire service and $19,900 from car sales. Present these data as a pie chart.
  (b) In the following three months petrol sales were $42,900 out of a total of $87,500. Represent this as an exploded pie chart.
  (c) By how much, as a proportion of the whole earnings, did petrol sales increase between the first and the second three-month periods?
(4) What are the essential differences between a column (or block) graph and a histogram.
(5) (a) From the data in question 1 develop the two moving average graphs.
  (b) Insert a trend line into each graph if appropriate.
(6) For $-4 \leq x \leq 4$ calculate and plot on a graph the function

$$y = 2x^2 + 2x + 4.$$

(7) (a) On a single graph plot the functions $y = x^3 + 4$ and $y = 27 - 2x$.
  (b) At what values of 'x' and 'y' do the functions intersect?
  (c) For $0 \leq x \leq 3$ and $y \geq 0$ show, by shading, the feasibility region for $y \leq x^3 + 4$ and $y \leq 27 - 2x$.
(8) A small manufacturer of steel boxes needs to produce no more than he can sell, or than he can store until the product *is* actually sold. His box-making machine cans be set up to produce *either* large or small boxes at any one time but not both.

The available storage space totals 4 m^3; large boxes occupy 1/100 m^2 whereas small boxes take up 0.008m^3. He can sell, at most, 280 large, or 400 small boxes per day and in each 8-hour day the machine can produce either large boxes at 40 per hour, or small boxes at 80 per hour.

(a) Construct a model of each of the restricting or constraining functions, for example: $4 \geq 0.01L + 0.0085S$.

(b) Calculate the three inequalities for an appropriate range of quantities of large and small boxes.

(c) Plot the three functions on a graph and outline the region which identifies the possible allocations of production between large and small boxes.

9
Matrices and graphics

Objectives

After working through this chapter you should be able to:

❑ relate shapes and their movements in space to the computer
❑ use elementary vectors
❑ understand the motion and transforms of two-dimensional (2D) shapes
❑ model shapes and operations on them by matrix arithmetic
❑ perform a range of elementary 2D spatial operations.

9.1 The new graphics

9.1.1 To anyone who is only used to traditional data processing a first encounter with desk-top publishing (DTP), computer-aided design (CAD) or computer-aided manufacture (CAM), much less computerized navigation systems or automatic pilots, may come as a great shock. How can the computer manipulate *spatial* information of such complexity and at such speed, or cause metal to be cut with such accuracy? All these things are based on **spatial models**.

9.1.2 The mathematical power of the graph, the graphical capabilities of the VDU screen and the modern printer or plotter, combine to create some of the most widely useful ranges of computer applications.

We are able to emulate and to manipulate entities in two- or three-dimensional (2D or 3D) space and to model real objects and their relationships in spatial terms. Eventually these same basic ideas extend to controlling, on the one hand the machine tools which cut or weld metal, on the other, the fly-by-wire aircraft. Essentially it is the computer programmer who plans, through the program(s), and thereafter controls, *movement in 3D space and time,* whether it be of the pencil, the aircraft, or the cutting edge of a machine tool.

9.1.3 As we have already seen the essential *spatial* modelling tool is the **Cartesian space**, a square grid on which the precise location of any point is determined by identifying an **ordered pair** of which the 'plus' and 'minus' signs then indicate the quadrant. (Note that the four quadrants give four possible sequences of $\pm x$, $\pm y$. The quadrant numbered 1 in Figure 9.1 is sometimes referred to as the '+,+' quadrant whereas that numbered 3 is the '−, −' quadrant and so on.)

Figure 9.1   The four-quadrant Cartesian space.

9.1.4 In the example in Figure 9.2 the point A, the intersection (at the centre of the circle, which is used as an *indicator* only), is defined, without any

Figure 9.2   Points on the grid.

ambiguity, by an ordered pair of the general type $(\pm x, \pm y)$, with the x value, or the horizontal offset from zero, always given first and followed by the y component, the vertical offset from zero. Point A is (7,3), point B is (–6,2), point C is (–5,–4) and D is (3,–3). (In this notation the common practice is to *assume that numbers are positive* unless otherwise shown and the '+' signs are omitted. Until you are very familiar with the notation you might find it wise to *include* the positive signs as a reminder to yourself.)

9.2 Vectors

9.2.1 In Figure 9.3 we see that the 'journey' from A to B may be indicated by a straight line, which, however, needs an arrow to make clear that it is from A to B and *not* from B to A. Such a line indicates a **change of**, or **difference in**, **position** and, (since a **difference** is the result of a *subtraction*), may be found by subtracting the first ordered pair, the starting position, from the second such pair, that of the finishing position; $(-6,2) - (7,3) = (-13,-1)$.

Figure 9.3   The journey from A to B – a vector.

9.2.2 The difference between any two points (or any quantum of which we know both magnitude and direction), may be indicated by such a **vector** which, in the present case (and as we saw in Chapter 7), also identifies a continuous linear function of **slope** or **ratio** –1:13, which has, in the *opposite direction*, $B \rightarrow A$, the *inverse slope* of 1:13. This tells us that for a change of 13 on the x axis the change on the y axis is 1.

9.3 Vectors and shapes

9.3.1 If we link *all* the points in sequence, A to B, B to C, C to D, D to A, with a succession of similar vectors, as is done in some elementary programming languages such as LOGO, we find that we have described a *shape*, in this case an *irregular quadrilateral*.

In narrative form we could say, 'If we start at the point (7,3), move 13 left and down 1; 1 right and down 6; right 8 and up 1; right 4 and up 6; we describe the shape **ABCD**.' Notice how plus equates with 'move to the right' and minus is 'move to the left' along the x axis but 'up' and 'down', respectively on the y axis.

9.3.2 These descriptions of movement or change of position (and the notation) are of great importance and we need to practise using them in a variety of situations.

Figure 9.4 A shape as a sequence of vectors.

9.3.3 This 'shape' also describes a *journey* from A to B to C to D and back to the start at A with the four 'legs' of the trip as the vectors:

$$\overline{AB}, \ \overline{BC}, \ \overline{CD}, \ \overline{DA}$$

9.4 Notation

9.4.1 In this notation we see that the *sequence* of letters shows the *direction* whilst the line *above* the letters indicates a **vector**, whose direction and length are both known (rather than a **line segment**, where length only is known, or a **ray**, of which we know the direction only). Both the concept of the vector and the notation are of great importance in many applications.

9.5 Combining vectors

9.5.1 We may also **combine vectors**. In our example the journey from A to B and then to C is equivalent, in terms of overall progress, to going *straight* from A to C. In vector terms

$$\overline{AB} \ + \ \overline{BC} \ = \ \overline{AC}$$

Let us see if the arithmetic works out in ordered pairs.

A to B = (−13,−1) *and* (or +) B to C = (1,−6) and adding these together we have (−12,−7). A direct path from A to C shows the *difference* of (−5,−4) − (7,3) which is, perhaps unsurprisingly, (−12,−7). The result of combining two or more vectors like this we call the **resultant**, again a very useful little tool.

9.5.2 You may well have worked such practical examples, for instance, finding the actual path of a boat when a rower can achieve, in still water, a speed of 6 km/h, but the river flows at 8 km/h. In the graphical model of the rower attempting to travel straight across the river, in Figure 9.5, the river is envisaged as flowing from the bottom of the graph to the top with the 'near' bank along the y axis.

The rower, starting at (0,0), could if there were no current, row straight across from A to B, represented by the vector AB with an arrow indicating direction. (The vector is shown here as a thick line for clarity of illustration only since a vector normally *has no thickness*.)

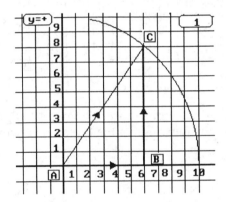

Figure 9.5   Modelling the river.

9.5.3 The current, flowing from B to C is also shown as a (thick line) vector. The result of combining *both* the speeds *and* the directions is the vector AC, known as the **resultant**, which shows that the boat would actually progress downstream at the angle CAB, rather than straight across the river.

By striking the arc of a circle from C to the horizontal scale we can also determine that the resultant *speed* would be 10 km/h. Where would the rower have to 'aim' his boat and how fast would he have to row in order to make his boat travel straight across the river at 6 km/h?

What would happen if the rower were to row faster, or slower? Try some variations on the original vectors and plot the resultants; as always your work should include trying several such examples until you are confident that you understand.

9.6 The numerical model

9.6.1 The final modelling tool in such applications is the **matrix** containing the **numerical model**, or set of ordered pairs, which may then be manipulated by well tried and proven mathematical techniques within existing laws.

9.6.2 To revert to a shape model (and we will confine ourselves to two dimensions only), in Figure 9.6 is a simple shape which is, quite deliberately, non-symmetrical.

It is five-sided and so our matrix representation of it will have five ordered pairs.

$$A = \begin{bmatrix} 0 & 3 & 5 & 4 & 1 \\ 0 & 2 & 2 & 4 & 3 \end{bmatrix}$$

9.6.3 We have, in this way, modelled our irregular pentagon in a form which is a very economical data structure and which we can operate upon by conventional matrix arithmetic.

Figure 9.6   The five-sided shape for exploration.

9.6.4 But what kinds of operation may we need to perform? In the real world objects are not static but dynamic. They move, grow, shrink and change their appearance in ways which traditional geometry does not envisage.

9.7 Transformations

9.7.1 It is not surprising that there is a **geometry** of **transformations**, part of **motion geometry**, in which we examine and operate upon shapes which are *dynamic*. The computer, more than any other of man's inventions, is a wonderful tool for modelling the dynamic world.

9.7.2 Let us start with some very simple changes.

9.7.3 In Figure 9.7 we see an **enlargement**: our original pentagon, *A* in this diagram, is enlarged to become *A'* which is clearly the same shape although of different size. Mathematically we identify this as **similar**, the shape is identical but the size is different.

Our original matrix:

$$A = \begin{bmatrix} 0 & 3 & 5 & 4 & 1 \\ 0 & 2 & 2 & 4 & 3 \end{bmatrix} \text{ has become } A' = \begin{bmatrix} 0 & 6 & 10 & 8 & 2 \\ 0 & 4 & 4 & 8 & 6 \end{bmatrix}$$

9.7.4 What is the mapping function which has converted *A* to *A'*? It does not take a great deal of investigation to reveal that all the pairs which make up *A* have been multiplied by 2 to generate *A'*.

Figure 9.7 The shape enlarged.

9.7.5 Such an enlargement, in matrix mathematics, is known as a 'scaling' and the multiplier is described as a **scalar**. We write the operation as:

$$2 * \begin{bmatrix} 0 & 3 & 5 & 4 & 1 \\ 0 & 2 & 2 & 4 & 3 \end{bmatrix} = \begin{bmatrix} 0 & 6 & 10 & 8 & 1 \\ 0 & 4 & 4 & 8 & 3 \end{bmatrix}$$

What happens if our scalar is ½? Try various *other* scalings and see what happens. (Make sure that the scale of your graph allows you to see clearly what is taking place.)

9.7.6 We describe all variations on an original shape, or its matrix, as **transforms** and the next such 'family' to consider are the **translations**. In a *translation* from one language to another we 'transfer' a piece of spoken or written text from the one language to the other *without changing its sense.*

9.7.7 A *mathematical* translation is very much the same; the shape is transferred to a new position without changing its appearance in any way.

9.7.8 Let us look first at a simple *lateral translation* in which our shape is merely 'slid' sideways.
Our origina matrix

$$A = \begin{bmatrix} 0 & 3 & 5 & 4 & 1 \\ 0 & 2 & 2 & 4 & 3 \end{bmatrix}$$

becomes

$$A' = \begin{bmatrix} 6 & 9 & 11 & 10 & 7 \\ 0 & 2 & 2 & 4 & 3 \end{bmatrix}$$

and, in comparing the two to see what *difference there is* between them, by *subtracting* the first matrix from the second, we find the matrix (with Δ as the symbol for *difference*), in which we have five repetitions of the vector (6,0)

$$\Delta = \begin{bmatrix} 6 & 6 & 6 & 6 & 6 \\ 0 & 0 & 0 & 0 & 0 \end{bmatrix}$$

9.7.9 We can, therefore, identify the transform A to A' in Figure 9.8 as

$$\begin{pmatrix} 6 \\ 0 \end{pmatrix} + \begin{bmatrix} 6 & 3 & 5 & 4 & 1 \\ 0 & 2 & 2 & 4 & 3 \end{bmatrix}$$

in which I have used the curved brackets to indicate the **column vector** which is to be added to each **column** of the matrix in turn to effect the transform.

Figure 9.8 A lateral translation.

9.7.10 There is, however, a **law of identity** for vectors and matrices which says that, 'if two or more vectors or matrices are identical *they are the same vector, or matrix.*' For an illustration of this look at the transform in Figure 9.9. For the movement of each corner of the shape to its new position check the vector by subtracting each starting point from its relative finishing point.

Figure 9.9 The common vector of a translation.

9.7.11 We can now look at various other translations in Figure 9.10 and consider what **column vector addition** generates the translations $A \to A'$, $A \to B$, $A \to C$ and $A \to D$, respectively. In each case you should be able to identify a single vector.

Can you see how the transform $A \to C$ is the same as ($A \to A'$ and then (+) $A \to B$)? How does this compare with our earlier combining of the 'journey' vectors when moving round the outline of a shape?

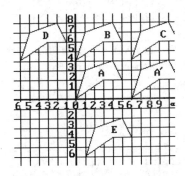

Figure 9.10 A variety of translations.

9.7.12 It is also possible to combine *transforms* to produce particular effects. In the case of CAD software, for example, we often find the functions which allow us to present our basic designs in *perspective*. This involves both a translation and an enlargement or diminution of the original shape, a very simple example is shown here. The original square has been reduced in size *and* translated *'left and up'* in order to help generate the simple perspective representation which we see in Figure 9.11.

Figure 9.11 A perspective view of a square prism.

9.7.13 What sequence of transforms would you use in order to achieve this? Is the sequence of operations important?

Try a number of such variations for yourself; sketching them in pencil on graph paper is the easiest way, but *do* remember to set them out carefully and *do* set up the necessary matrix and the transform scalars and vectors to complete the operations on the *numeric* models.

There are many such combinations which we use in CAD and similar graphics software and we shall meet others.

9.8 Reflections

9.8.1 Yet another family of transforms are the **reflections**, in which the shape is converted into its own mirror image, in the *y* axis, in the *x* axis, or *both*. In the illustration at Figure 9.12 on the next page we can see the three possible reflections of our original shape, *A*, into *A'*, *B* and *C*, respectively.

9.8.2 What is shape 'D'?

9.8.3 Compared with our original matrix the first reflection, into *A'*, is the matrix

$$\begin{bmatrix} 0 & -3 & -5 & -4 & -1 \\ 0 & 2 & 2 & 4 & 3 \end{bmatrix}$$

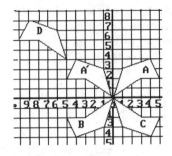

Figure 9.12   The possible reflections.

in which all the *x* values, in the top row, have become *negative* (−0 does not exist, there is only 0).

9.8.4 No scalar or vector multiplication can achieve this result and it is time for us to meet both the methods of matrix multiplication *and* a rather unusual matrix.

9.9 Matrix multiplication

9.9.1 In matrix multiplication there are a number of special rules and we need a little extra practice to become familiar with them.

The first rule is that the number of *columns* in the multiplier *must* match the number of *rows* in the multiplicand. The second is rather more complicated and needs to be taken in stages. Taking the top row of the *multiplier* first, each of its elements, from left to right in turn, must multiply, in strict sequence, each element, from top to bottom, of the first column of the multiplicand matrix and the partial products must be totalled to give the *top left* cell of the product matrix.

The top row of the multiplier must then multiply the *second* column of the multiplicand, in the same fashion, to generate the *second column cell* in the *top row* of the product matrix. This process continues until all the columns, from left to right, of the multiplicand, have been multiplied by the top row of the multiplier. We then repeat the process with the *second* row of the multiplier, starting again with the leftmost column of the multiplicand, to furnish the *second row of the product matrix*. This process continues until all row elements of the multiplier have multiplied all column elements of the multiplicand and we have a complete product matrix.

This is a rather involved process to explain and can be quite a tedious process to perform on a large matrix, manually, but a program segment to perform it is surprisingly brief. Let us explore the stages by using two matrices

9.9.2 Under the first rule

$$
\begin{bmatrix} 1 & 2 & 3 \\ 0 & 2 & 1 \end{bmatrix} * \begin{bmatrix} 2 & 1 & 0 \\ 1 & 4 & 3 \\ 2 & 1 & 5 \end{bmatrix}
$$

is legitimate but

$$
\begin{bmatrix} 2 & 1 & 0 \\ 1 & 4 & 3 \\ 2 & 1 & 5 \end{bmatrix} * \begin{bmatrix} 1 & 2 & 3 \\ 0 & 2 & 1 \end{bmatrix}
$$

is not since the columns in the multiplier do not match the rows of the multiplicand.

9.9.3 If the multiplier

$$
\begin{bmatrix} 1 & 2 & 3 \\ 0 & 2 & 1 \end{bmatrix}
$$

designated as *A*, the multiplicand

$$
\begin{bmatrix} 2 & 1 & 0 \\ 1 & 4 & 3 \\ 2 & 1 & 5 \end{bmatrix}
$$

labelled *B* and the product matrix *P*, let us 'step through' $A * B = P$, by the rules of matrix multiplication.

Step 1 is:

$$
\begin{bmatrix} 1 & 2 & 3 \\ \cdot & \cdot & \cdot \end{bmatrix} * \begin{bmatrix} 2 & \cdot & \cdot \\ 1 & \cdot & \cdot \\ 2 & \cdot & \cdot \end{bmatrix} = \left| 1*2+2*1+3*2 = 10 \right|
$$

to form the first cell in the product matrix:

$$
\begin{bmatrix} 10 & \cdot & \cdot \\ \cdot & \cdot & \cdot \end{bmatrix}
$$

The next step is to multiply *column two by row one*

$$\begin{bmatrix} 1 & 2 & 3 \\ \cdot & \cdot & \cdot \end{bmatrix} * \begin{bmatrix} \cdot & 1 & \cdot \\ \cdot & 4 & \cdot \\ \cdot & 1 & \cdot \end{bmatrix} = \left| 1*2+2*4+3*2 = 12 \right|$$

to form cell $p_{1,2}$ in our product matrix, P:

$$\begin{bmatrix} 10 & 12 & \cdot \\ \cdot & \cdot & \cdot \end{bmatrix}$$

Next, row three performs:

$$\begin{bmatrix} 1 & 2 & 3 \\ \cdot & \cdot & \cdot \end{bmatrix} * \begin{bmatrix} \cdot & \cdot & 0 \\ \cdot & \cdot & 3 \\ \cdot & \cdot & 5 \end{bmatrix} = \left| 1*0+2*3+3*5 = 21 \right|$$

to form cell $p_{1,3}$ in our product matrix:

$$\begin{bmatrix} 10 & 12 & 21 \\ \cdot & \cdot & \cdot \end{bmatrix}$$

We now move to the second row of our multiplier and start with

$$\begin{bmatrix} \cdot & \cdot & \cdot \\ 0 & 2 & 1 \end{bmatrix} * \begin{bmatrix} 2 & \cdot & \cdot \\ 1 & \cdot & \cdot \\ 2 & \cdot & \cdot \end{bmatrix} = \left| 0*2+2*1+1*2 = 4 \right|$$

to form the first cell in the second row of the product matrix:

$$\begin{bmatrix} 10 & 12 & 21 \\ 4 & \cdot & \cdot \end{bmatrix}$$

and continue the process until we have

$$P = \begin{bmatrix} 10 & 12 & 21 \\ 4 & 9 & 11 \end{bmatrix}$$

9.9.4 As a pencil-and-paper process this is tedious for all but quite small matrices, but a program segment is relatively short. Here is one possible version in which merely changing the maxima for i, j and k will adapt the routine for any sizes of matrix, subject to the first rule, above. For brevity, I have dealt only with the 'core' activity and not declared the arrays, merely labelled them as in our worked example, above.

```
i = 1
  DOWHILE i < 3
  j = 1
    DOWHILE j < 4
    k = 1
      DOWHILE k < 4
      Pi,j ← Pi,j + (ai,k * bk,j)
      k ← k + 1
      ENDDO
    j ← j + 1
    ENDDO
  i ← i + 1
  ENDDO
END
```

9.9.5 Beware matrix multiplication is **not commutative!**
Try

$$\begin{bmatrix} 1 & 2 & 3 \\ 0 & 2 & 1 \end{bmatrix} * \begin{bmatrix} 1 & 0 \\ 4 & 3 \\ 1 & 5 \end{bmatrix}$$

and then try

$$\begin{bmatrix} 1 & 0 \\ 4 & 3 \\ 1 & 5 \end{bmatrix} * \begin{bmatrix} 1 & 2 & 3 \\ 0 & 2 & 1 \end{bmatrix}$$

Are the product matrices identical?

9.9.6 Let us now look at the special matrix which generates reflections. Because of the pattern of matrix multiplication we use many variations of the rather odd-looking matrix

$$\begin{bmatrix} n & 0 \\ 0 & n \end{bmatrix}$$

as a multiplier which *will multiply selectively*.

9.9.7 The rules are that this matrix must be a *square* with the same number of *columns* as there are *rows* in the matrix which is to be multiplied and that all cells not on the **leading diagonal** shall be zero.

The leading diagonal is that from top left to bottom right and we replace the n of the model, above, with whatever we need in order to achieve the required result.

9.9.8 Most importantly the **identity matrix**, with one in each cell of the leading diagonal and zero in all others, will leave the *values unchanged* in any matrix which it multiplies.

In our present case, for example

$$\begin{bmatrix} 1 & 0 \\ 0 & 1 \end{bmatrix} * \begin{bmatrix} 0 & 3 & 5 & 4 & 1 \\ 0 & 2 & 2 & 4 & 3 \end{bmatrix} * \begin{bmatrix} 0 & 3 & 5 & 4 & 1 \\ 0 & 2 & 2 & 4 & 3 \end{bmatrix}$$

What happens if we make:

(a)   the top left cell –1?
(b)   the bottom right cell –1?
(c)   *Both* cells on the leading diagonal –1?

What combination caused $A \rightarrow D$ in Figure 9.12?

9.9.9 This surprisingly simple device allows us to do some very sophisticated transforms and we will see other variations a little later.

9.10 Rotations

9.10.1 For the moment let us look at another major set of transforms, the **rotations**. In Figure 9.13 we see our shape rotated to various extents until, finally, it can turn a little further from C and become 'itself' again. In other words a 360° rotation is achieved by the **identity** matrix once again.

Figure 9.13   Four of an infinite number of rotations.

9.10.2 This time, however, we are concerned with the *amount of turn*, or the angle through which we turn our shape, as in the **rotate** function of many graphics programs.

9.10.3 The matrix which achieves the rotational transforms is

$$\begin{bmatrix} \cos\alpha & -\sin\alpha \\ \sin\alpha & \cos\alpha \end{bmatrix}$$

where α denotes the size of angle of *anticlockwise* rotation which we seek.

9.10.4 It is perhaps not surprising that the transform matrix, for this purpose, contains **functions of the angle** of turn. It need not concern us how the operative matrix is arrived at and the angular functions are normally available in most interpreters and compilers.

The only complication is that few such translators deal readily with the 60-based fraction system of degrees, minutes and seconds with which we customarily measure angles but use, instead, the decimal-based **radian measure**.

Perhaps we should pause for a moment and clarify what this is. Figure 9.14 may help.

Approx. 57.2958°
1 RADIAN

Figure 9.14 One radian = $360°/2\pi$.

9.10.5 A distance, *equal to the radius, along the circumference of the circle*, marks off an **arc** which defines an angle at the centre of the circle of **one radian**. Since the circumference is $2\pi r$, or $2 * \pi *$ the length of the radius, one radian $= 360°/2\pi$ which is approximately $57.2958°$. This is a **constant** and to convert angles expressed in degrees to radian measure we divide by this constant.

Be sure to check carefully, in all computing work, whether *your* translator accepts angles in degrees or radians; you may need to write a conversion program module.

9.11 Glide and stretch

9.11.1 When we perform both a rotation and a translation we describe it as a **glide**. There is, as with all our transforms, no real way of learning about such behaviour except by having plenty of practice and you should try as many examples as possible with different shapes, different amounts of turn and different combinations of transform.

9.11.2 So far all our transforms have left the original shape *unchanged* but let us look at other variants of the identity matrix and see what happens if we use values other than one in the leading diagonal

$$\begin{bmatrix} 2 & 0 \\ 0 & 1 \end{bmatrix} * \begin{bmatrix} 0 & 3 & 5 & 4 & 1 \\ 0 & 2 & 2 & 4 & 3 \end{bmatrix} = \begin{bmatrix} 0 & 6 & 10 & 8 & 2 \\ 0 & 2 & 2 & 4 & 3 \end{bmatrix}$$

Plot this on your graph paper; how would you describe what has happened to our original shape?

9.11.3 This is a **stretch**. In this case we have stretched our shape along the *x* axis but left the *y* values unchanged. What happens if we make both cells on the leading diagonal 2; or top left 2 and bottom right 1½? Clearly this is another transform which we need to investigate by experiment to be sure of what effect we shall produce with different values in the leading diagonal.

9.12 Shear

9.12.2 The final shape-changing set of transforms, for our present purpose, is achieved by yet another modification of the identity matrix. These perform an operation known as **shear** of which Figure 9.15 gives us just one example.

Figure 9.15   The shape distorted – but still 'true'.

9.12.3 By using the transform matrix:

$$\begin{bmatrix} 1 & 2 \\ 0 & 1 \end{bmatrix}$$

on our shape, A, we arrive at the product matrix, A'

$$\begin{bmatrix} 0 & 7 & 9 & 12 & 7 \\ 0 & 2 & 2 & 4 & 3 \end{bmatrix}$$

and when this is plotted on our graph field we see the rather surprising result in Figure 9.15. It is as though a force pushed the shape, at the top-left corner, so that every part of the shape 'slides' sideways by a different amount.

These comparatively simple models, and the operations upon them, are equally well performed in three dimensions, by using ordered triples and three-row matrices. They *are* a little beyond the scope of this book, but the principles are precisely the same and many of the most exciting developments in computer applications depend upon just such manipulations.

9.13 Conclusion

The graphical capabilities of the computer, at least in *vector* graphics, are really *numeric* capabilities relying on *our understanding* of the underlying mathematics in order to switch our thinking from graphic to numeric mode at will. The screen and the plotter are, ultimately, Cartesian number spaces in common with our everyday graphs.

Exercises

Use graph paper to represent a VDU screen wherever necessary.

(1) An aircraft is flying due south at 400 knots and there is a 50-knot wind blowing from the east. Plot the vectors and show the plane's actual track.

(2) A boat sails 5 km due south, then 7 km north-east followed by 3 km due north.

   (a) Using the *y* axis as north/south and a scale of 1 cm:1 km plot the boat's track and measure the distance and bearing it needs to sail in order to get back to its starting point.

   (b) State this vector as an ordered pair.

(3) *S* is approximately 200 km east and 200 km south of *K* and *I* is approximately 50 km west and 150 km north of *K*. What is the vector from *I* to *S*?

(4) A shape is identified by the following matrix:

$$S = \begin{bmatrix} 1 & 5 & 1 & 2 & 0 \\ 2 & 1 & 1 & 5 & 2 \end{bmatrix}$$

Plot the shape on a graph.

(5) In each of the following operate on *S*, as in (3), and plot the change on graph paper (or if you have a suitable computer system write the appro priate short program and test it).

   (a) What is the result of

$$\begin{pmatrix} 2 \\ 3 \end{pmatrix} + S?$$

   (b) Calculate

$$\begin{bmatrix} -1 & 0 \\ 0 & 1 \end{bmatrix} * S$$

   (i) Plot the resultant shape.
   (ii) What transform has been effected?

   (c) What is the effect on the original shape of 1.5 * *S*?

   (d)

$$\begin{bmatrix} 0 & -1 \\ 1 & 0 \end{bmatrix} * S$$

   (e)

$$\begin{bmatrix} -1 & 0 \\ 0 & -1 \end{bmatrix} * \left(\begin{pmatrix} 0 \\ 2 \end{pmatrix} + S \right)$$

(f)
$$\begin{bmatrix} 1 & 1.5 \\ 0 & 1 \end{bmatrix} * \left(\begin{bmatrix} -1 & 0 \\ 0 & 1 \end{bmatrix} * S \right)$$

(6) Write and test a pseudo-code segment to perform each of the tasks in **(5)**.

(7) **(a)** A shape is created by joining the points (−5,−1), (−2,0) and (−3,2). Write down the matrix *A* which fully represents this shape.

(b) The following matrices

$$T1 = \begin{bmatrix} 8 & 8 & 8 \\ 1 & 1 & 1 \end{bmatrix} \text{ with } T2 = \begin{bmatrix} 0 & -1 \\ 1 & 0 \end{bmatrix}$$

are used to operate on *A*. Find *B* and *C* when *B* = *T*1 + 2 * *A* and *C* = *T*2 * *A*.

(c) Describe the transforms *B* and *C*.

(8) What transform matrix is needed to move any shape 3 units to the left and 4 units 'up' on a grid?

10

Data structures

Objectives

After working through this chapter you should be able to:

❑ understand the relevance and operation of arrays, lists, stacks, queues, trees and similar elementary data structures
❑ program the loading, searching and ordering of data structures
❑ describe how networks are stored and manipulated as arrays
❑ understand and use tables with pointers and back-pointers
❑ select the structure which is appropriate to the task.

10.1 Organizing data

10.1.1 While the computer is capable of very fast processing it completely lacks imagination or any ability to 'think' for itself. One consequence of this is that it is very difficult to write programs to process data which are presented or stored haphazardly. It is also true that the most *efficient* processing can be attained when data are organized in ways which suit the machine particularly well. Once again, mathematics offers ready-made solutions to such problems.

10.1.2 The first thing which we demand of good storage and processing is that the data be organized according to an efficient *pattern* so that we do not need a separate memory address for each data item. Ideally, the nature of the pattern should be defined *simply*, either within the program or by means of a *header* within the data set itself, perhaps as the first record, or as a 'flag' to that record.

10.1.3 The second need is that we should be able to find any *individual* piece of data quickly. Some arrangements allow faster searches than others.

10.1.4 Thirdly we need to be able to insert, exchange or delete individual data items readily.

10.1.5 It is the acceptable *practical* **data structures** which match these needs which we shall begin to study in this chapter. We shall look at some of the basic data structures and examine how they are used and what elementary manipulations may be performed on them.

10.2 Matrices again

10.2.1 Since so much of the data we are handling is *numeric* it is not surprising that the **matrix** or **array** is one of the commonest data structures. We have already looked at the basic principles of matrices and seen some applications. For most purposes we commonly process arrays by 'nested loops' in our programming.

10.2.2 To load the data into an array, for example, we might first *declare* the array by **(a)** defining the data type and **(b)** *dimensioning the array*. We may for example, make two program statements such as DEFINT A, to declare A as an integer, and DIM A(2,5), to dimension the array A as a two-row by five-column matrix.

10.2.3 We are not tied to *two* dimensions but may create one-dimensional arrays (vectors), or define several dimensions; the computer is not limited in this respect.

10.2.4 To input a data set into the matrix, A, we might then use a segment of 'nested loop' code rather like this:

```
i = 1, j = 1;
DOWHILE i < 3
   DOWHILE j < 6
   READ A_{i,j}
   j ← j + 1
   ENDDO
   j ← 1, i ← i + 1
ENDDO
END
```

which will read successive inputs into $a_{1,1}, a_{1,2}, ..., a_{1,5}$ in the first 'pass' of the i loop, involving *five* passes of the j loop. On the second pass of the i loop cells $a_{2,1}, a_{2,2}, ... , a_{2,5}$ will be 'loaded'.

10.2.5 The same program segment can be adapted to perform any operation normally performed on an array. Given the speed of processors this is a perfectly adequate mechanism for the vast majority of array processing but, for some highly specialized purposes, **array processors** may be installed. In machines which are so configured each cell, of perhaps an 8 × 8 array, is allocated its own dedicated processor so that all cells may be processed simultaneously (rather than serially, as in our nested loop structure).

Data sets which are larger than 8 × 8 can then be subdivided into a sequence of arrays of that size for processing.

10.2.6 It is common for spreadsheet systems to permit any two or more subsections of the sheet to be defined as arrays and the normal operations of add, subtract, multiply and divide can be called by fixed commands in the spreadsheet meta-language or macros. Since the commutative and inverse laws do not operate in the same way for matrices *we* must specify not only the arrays to be used and an output array but, most importantly, the *sequence* of arrays in the calculation.

10.2.7 The common operations on a matrix, some of which we have already met in other contexts, are:

(1)   It may have a constant quantity *or* a constant vector, added to it or subtracted from it.

(2)   It may be multiplied, or divided, by a (constant) *scalar*.

(3)   It may have another matrix, *of the same dimensions*, added to it or subtracted from it by *cell-by-cell mapping*.

(4)   It may, if we specify the procedure clearly, be subjected to a *cell-by-cell* multiplication by another matrix of *identical dimensions*. This is *not true* matrix multiplication and cannot usually be achieved by calling such a function. [Within a spreadsheet performing a formula multiplication of the corresponding (1,1) cells and then using the **copy** function is often the answer. Within a written program the nested loop construct will serve.]

(5)   It may be multiplied by a *row vector* which has the same number of elements as the original matrix has rows or it may multiply a *column vector* which has the same number of rows as the original matrix has columns.

(6)   It may be multiplied, selectively, by a *square matrix* of which the number of columns matches the number of rows in the multiplicand.

(7)   It may be multiplied, (in *true* matrix multiplication, sometimes described as 'row-by-column'), by a matrix, *the number of whose columns match the number of rows of the multiplicand matrix*. For example a 4 × 6 matrix, in 'full' matrix multiplication, may be multiplied by any $n \times 4$ matrix where $1 \leq n \leq 6$.

10.3 An array example

10.3.1 Let us take a practical example and consider in turn, each of these processes which we have not previously met. We first need the 'narrative' of the problem which we must model. "A manufacturing company, in factories at Andover and Bristol, respectively, employs the following numbers of people in four grades for each of six job-categories and at different monthly pay scales. (For simplicity the data are given in the tables.) There are a number of questions to which management need answers and other such questions will arise in the future.'

10.3.2 Table 10.1 shows the number of staff in each of the two factories set out in rows and columns and this already constitutes a matrix of a sort but it is not yet very suitable for manipulations. For that we must simplify yet further.

Table 10.1

	Andover						Bristol					
	Category						Category					
Grade	A	B	C	D	E	F	A	B	C	D	E	F
1	0	2	5	16	0	0	0	3	18	0	0	1
2	2	12	12	27	5	1	1	4	12	18	2	0
3	0	0	24	15	3	2	0	0	16	14	5	1
4	0	0	10	9	2	1	0	0	8	6	1	0

10.3.3 In Table 10.2 we have the monthly pay *rates* set out and for the wages element of costs, if we organize these three sets of data into the appropriate matrices, we can derive a good deal of further information by using, at different stages, some of the operations listed as **(1)** to **(7)**, above.

Table 10.2

	Wage rates $00					
	Category					
Scale	A	B	C	D	E	F
1	1	1	1	1	2	2
2	2	2	3	3	4	5
3	3	3	4	4	6	8
4	8	8	9	9	10	12

10.3.4 The first step in modelling the situation is to create true matrices from the known data with all rows and columns labelled appropriately. It does not matter particularly which items we put into the rows and which into the columns so long as, thereafter, we are wholly consistent.

If A = Andover, B = Bristol and W = Wage rates then:

$$A = \begin{bmatrix} 0 & 2 & 5 & 16 & 0 & 0 \\ 2 & 12 & 15 & 27 & 5 & 1 \\ 0 & 0 & 24 & 15 & 3 & 2 \\ 0 & 0 & 10 & 9 & 2 & 1 \end{bmatrix}, B = \begin{bmatrix} 0 & 3 & 18 & 0 & 0 & 1 \\ 1 & 4 & 12 & 18 & 2 & 0 \\ 0 & 0 & 16 & 14 & 5 & 1 \\ 0 & 0 & 8 & 6 & 1 & 0 \end{bmatrix}$$

$$C = \begin{bmatrix} 1 & 1 & 1 & 1 & 2 & 2 \\ 2 & 2 & 3 & 3 & 4 & 5 \\ 3 & 3 & 4 & 4 & 6 & 8 \\ 8 & 8 & 9 & 9 & 10 & 12 \end{bmatrix}$$

contain all the known data and may now be manipulated in order to extract information.

10.3.5 Question: How many people *in all* are employed in *each* of the 24 sub-groups?

10.3.6 Method: Add the two 'numbers' matrices, A and B, cell-by-cell, to give T, the totals by category/grade.

$$\begin{bmatrix} 0 & 2 & 5 & 16 & 0 & 0 \\ 2 & 12 & 15 & 27 & 5 & 1 \\ 0 & 0 & 24 & 15 & 3 & 2 \\ 0 & 0 & 10 & 9 & 2 & 1 \end{bmatrix} + \begin{bmatrix} 0 & 3 & 18 & 0 & 0 & 1 \\ 1 & 4 & 12 & 18 & 2 & 0 \\ 0 & 0 & 16 & 14 & 5 & 1 \\ 0 & 0 & 8 & 6 & 1 & 0 \end{bmatrix} = \begin{bmatrix} 0 & 5 & 23 & 16 & 0 & 1 \\ 3 & 16 & 27 & 45 & 7 & 1 \\ 0 & 0 & 40 & 29 & 8 & 3 \\ 0 & 0 & 18 & 15 & 3 & 1 \end{bmatrix}$$

is what $A + B = T$ involves and is arrived at by the

$$a_{1,1} + b_{1,1} = t_{1,1}, ..., a_{4,6} + b_{4,6} = t_{4,6}$$

sequence in a DOWHILE i < 5; DOWHILE j < 7 nested loop. Try to write the pseudo-code routine for this, remembering to declare and initialize all necessary matrices and variables.

10.3.7 Question: how many people are employed, *in total*, at Andover and at Bristol?

10.3.8 Method: *Step 1*. Multiply *T* by the appropriate **unit row vector**. (A unit row vector is a single row matrix which has a 1 to match each **row** of the multiplicand matrix. The subsequent 'multiply by 1' process is using the law of identity to select and total cells without changing the individual values):

$$\begin{pmatrix} 1 & 1 & 1 & 1 \end{pmatrix} * \begin{bmatrix} 0 & 2 & 5 & 16 & 0 & 0 \\ 2 & 12 & 15 & 27 & 5 & 1 \\ 0 & 0 & 24 & 15 & 3 & 2 \\ 0 & 0 & 10 & 9 & 2 & 1 \end{bmatrix} = \begin{pmatrix} 2 & 14 & 54 & 67 & 10 & 4 \end{pmatrix}$$

Step 2. Multiply the **unit column vector** (which has a 1 to match each column of the multiplier **row vector**), by the **row vector** generated in step 1

$$\begin{pmatrix} 2 & 14 & 54 & 67 & 10 & 4 \end{pmatrix} * \begin{pmatrix} 1 \\ 1 \\ 1 \\ 1 \\ 1 \\ 1 \end{pmatrix} = 151$$

Can you **(a)** perform the same calculation for Bristol, **(b)** for both factories *in total*, **(c)** write the pseudo-code segment which will perform this task?

10.3.9 Question: what is the total wages bill?

10.3.10 Method: *Step 1*. Perform the cell-by-cell addition, *A* + *B* to find the 'total employees' matrix, *T*.

Step 2. Next, do the *cell-by-cell* multiplication of *W* * *T* to generate the matrix which will contain, (by category and grade), all wages costs, *C*

$$C = \begin{bmatrix} 0 & 5 & 23 & 16 & 0 & 2 \\ 6 & 32 & 81 & 135 & 28 & 5 \\ 0 & 0 & 160 & 116 & 48 & 24 \\ 0 & 0 & 162 & 135 & 30 & 12 \end{bmatrix}$$

Step 3. Multiply *C* by the appropriate unit row matrix to total the columns. (This technique may be useful for answering additional questions. Can you suggest such a question?)

Step 4. Multiply the appropriate unit column vector by the product of step 3 to generate the overall total. If you have followed steps 3 and 4 correctly your answer should be $102,000.

10.3.11 Question:what is the total wage bill for employees in categories C to E, inclusive, at grades 2 and 3, of matrix C?

10.3.12 Method:

$$\begin{pmatrix} 0 & 1 & 1 & 0 \end{pmatrix} * \begin{bmatrix} 0 & 5 & 23 & 16 & 0 & 2 \\ 6 & 32 & 81 & 135 & 28 & 5 \\ 0 & 0 & 160 & 116 & 48 & 24 \\ 0 & 0 & 162 & 135 & 30 & 12 \end{bmatrix} * \begin{pmatrix} 0 \\ 1 \\ 1 \\ 1 \\ 1 \\ 0 \end{pmatrix}$$

It is important that you should work this example and others like it so that you can see clearly the *selective* operation of the unit vector, row *or* column, in which some of the ones have been replaced by zeros, putting into effect the law of multiplication by zero to eliminate unwanted partial products in the matrix multiplication.

10.3.13 Question: what will be the effect on the total wage bill if all grade 1 staff are given wage increases of 5%, grade 2, 6%; grade 3, 4% and grade 4, 8%?

10.3.14 Method: *Step 1.* Since multiplying by 1.05, for example, will increase an amount by 5% we use a *I*-identity matrix (as we saw in the previous chapter for *selective* enlargement of shapes), which will first multiply each *row* by its appropriate scalar)

$$\begin{bmatrix} 1.05 & 0 & 0 & 0 \\ 0 & 1.06 & 0 & 0 \\ 0 & 0 & 1.04 & 0 \\ 0 & 0 & 0 & 1.08 \end{bmatrix} * C = \begin{bmatrix} 0 & 5 & 23 & 16 & 0 & 2 \\ 6 & 32 & 81 & 135 & 28 & 5 \\ 0 & 0 & 160 & 116 & 48 & 24 \\ 0 & 0 & 162 & 135 & 30 & 12 \end{bmatrix}$$

and find the *new* total wages matrix C'. Work this yourself before moving to step 2 and again, note the effect of the multiplications by zero.

Step 2. Find the column totals by multiplying the new matrix by the appropriate unit row vector.

Step 3. Find the overall total by multiplying the appropriate unit column vector by the product of step 2.

How would you evaluate the total increase? If you answer is 5.94% to two decimal places your intermediate steps were probably sound.

10.4 Arrays and spreadsheets

10.4.1 It is especially easy, with computerized manipulation of arrays, to perform 'what if?' modelling of real situations. In our previous example 'what if the grade 3 people had a 6% increase?' means only changing that one cell in the transform matrix and re-running the short program. In such ways, using an algorithmic structure-handling process or processes we can nevertheless develop and use genuinely 'try it and see', heuristic, models.

10.4.2 All such capabilities are built into any reasonable spreadsheet system where we can identify any subsection of the spreadsheet (itself an array), as a separate matrix, by identifying the **range** of cells occupied by the selected matrix, on which we may operate by any of the techniques we have discussed here. For example, to avoid a lot of pencil and paper working with the risk of making mistakes, a spreadsheet package was used to perform all the calculations needed in Section 10.3.

10.4.3 Try out on a spreadsheet, for yourself, the various operations we have performed, above. You will probably be asked to identify, by *range*, matrix 1, the multiplier, then matrix 2, the multiplicand and finally the *destination* of the product matrix, as its top left-hand cell. You should try full multiplication, as well as multiplication:

❏ by a row vector; by a unit row vector; by a selective row vector
❏ of a column vector; of a unit column vector; of a selective column vector.

10.5 Multidimensional arrays

10.5.1 For practical purposes pencil-and-paper modelling is limited to fairly small two-dimensional arrays but spreadsheets offer us much more power than this and we can work with quite large arrays.

10.5.2 However, it is undoubtedly within a program that we can develop the greatest power in manipulating arrays. For instance, *we* may find it difficult even to envisage three-dimensional data structures but the computer can deal as readily with *six*-dimensional arrays if they suit our purpose. This may sound a little strange at first but if we are dealing with three-dimensional movement, of an aircraft let us say, we are also dealing with both speed and time and a five-dimensional data array will be extremely useful.

Both the array structure and the methods of manipulating data within the structure have many applications and are of very great importance.

10.6 String arrays

10.6.1 The computer gives us yet more power because, although we cannot perform arithmetic operations on them, we *can* store strings in arrays, as well as numeric data, and may well be able at least to 'concatenate' the elements of one array with the elements of another. The functions available will depend upon the programming language and the individual interpreter software.

10.6.2 Unless our strings are of very varied lengths the array is a very useful structure for their storage and handling, especially in main memory because of its economical use of space.

10.7 Array addresses

10.7.1 Every additional memory address which we use will itself both use up memory and slow the processing. We therefore need to economize on addressing wherever possible. In order to do this when we define an array it is normally only given a start address and a total length, usually in bytes, by the program compiler or interpreter. The address label is transparent to the user but many translators allow us to identify it if we wish.

10.7.2 If an array is defined as 4×7 integer, and the computer stores integers in 16-bit (two-byte), form, we can readily visualize the array. From the starting address the control unit will allocate a nominal $4 * 7 * 2 = 56$ bytes of memory. Of this, since it is identified as 4×7, the first quarter of the space, 14 bytes, will contain the cells, in order, of the top row of the matrix. The next quarter will contain the second row, and so on.

10.7.3 We need know only the *starting address* where the array is stored; all else is relative to this and we speak of 'relative addressing'. We would load our data into each of their appropriate cells by some variation of a nested loop 'read' program segment.

10.8 Lists

10.8.1 When we look at a *real* data item, say one record in an invoice file, we often find that a single matrix cannot hold all the elements because they are of different data types. Such a record will need at least four fields, **invoice number: customer number: date: amount**.

The first item might be alphanumeric, the second an integer, the third a date string of the *fixed* dd/mm/yy type whilst the fourth may be a real number.

10.8.2 We could not define a four row matrix to hold four different variable types such as these and would need to declare *a different one-dimensional* array for each different data type.

10.8.3 Any such array or vector may also be thought of as a **list** and lists may be **independent** or **linked**. In our example shown in Figure 10.1, all four elements of the data record must stay together. This is not always easy since we may well, for different reports, need to *order or 'sort'* each list individually.

Figure 10.1 Four lists which must stay 'linked'.

10.8.4 In this case, as in Figure 10.2, the individual elements will no longer necessarily be anywhere near their original positions and our four lists would need to be linked, laterally, so that any individual invoice number 'points' to the position in the customer number list where the second element of the whole record is to be found.

Similarly the customer number should point to the date and the date to the amount so that, under all circumstances, entering the invoice number will automatically allow us to retrieve all the other elements of the corresponding record.

10.8.5 In practice we might well want all pointers to operate in both directions so that, for instance we could retrieve all invoices billed on a given date, or for a particular customer number, or for a given range of values.

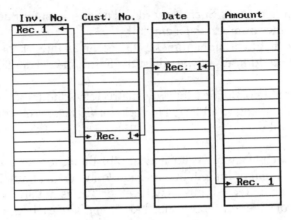

Figure 10.2   The lists after ordering.

10.8.6 The particular problem is that the *unique identifier* for each record is the invoice number, so that it is only this which establishes the combination of lists *as a set.*

10.8.7 At all costs we must preserve the mapping of the invoice number list onto each of the other three or we shall finish up with one set and three jumbled 'heaps'.

10.8.8 The easiest way, in our matrix techniques, of preserving the linking identity of each element of a number of lists simultaneously is to 'flag' each item by its *subscript* at the time of original entry, and to append this number to the data element as it is stored.

10.8.9 Such a device must be planned and provided for from the very beginning of program design and there is a strong case for including all such pointers in the data dictionary and all other documentation. The pointers may indicate one direction only, and a one-to-one mapping, or may be required to indicate a one-to-many mapping.

10.9 System lists

10.9.1 Program location in main memory and its control during execution is often by means of linked lists and Figure 10.3 is a sketch of how this might be achieved. Any single process or command will 'call' a particular segment of program code. Within, this in turn, may well be one or more calls to subroutines.

10.9.2 If each of the substructures of a single program, or the operating

Figure 10.3 Program lists with pointers.

system, is brought into memory as a list in a preset position we can, by setting suitable pointers, both economise on storage and speed up the running of the program.

10.9.3 We often achieve this by recording or 'poking' the relevant memory addresses into another list dedicated to the pointers themselves. This would entail yet another list added to those in Figure 10.3

10.9.4 Yet another use of lists is in holding in memory the various interrupt vectors which control the whole computer configuration while it is running.

10.9.5 Some understanding of the use of lists, their setting-up and operation does help us design better programs and understand what is going on when we read the programs of others.

10.10 Stacks

10.10.1 There are different ways of establishing and handling lists. If we try to read a manual on low-level programming, in assembler for example, we shall quickly meet the word 'stack' just as, in a job control language, or when managing a print spooler, we meet the word 'queue'. This applies to all levels of computing practice not merely to mainframe installations.

There is nothing to fear in these words since they, together with several others, are only labels for variants of one-dimensional arrays. What is different is the ways in which data are manipulated into and out of the arrays.

10.10.2 Any single item of data is, in strict mathematical language, a 'datum' and we shall start to use this word rather than repeating phrases like 'data item'. The other jargon words are 'push', for entering a datum into an array and 'pop' for retrieving a single datum.

10.10.3 If we pop and push data on a 'last-in-first-out' system we refer to such an array as a **stack**.

10.10.4 Figure 10.4 shows a graphic model of such a stack and there are a number of things to note. If we have allocated a stack for 16 data and, as yet, we have included only 5 of those data we shall neither know nor care exactly whereabouts in the stack they are.

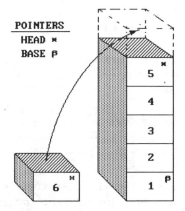

Figure 10.4   A stack of data 'packets'.

10.10.5 The control unit, however, *must* know. And so, though invisible to us, 'pointers' are set, usually in adjacent bytes of memory, which indicate exactly where both the first *and* the last data to be pushed into the stack are to be found.

10.10.6 If, as suggested in Figure 10.4, another datum is then pushed into the stack it will be set 'on top of' datum 5 and the stack 'top' pointer will be reset to indicate datum number 6.

10.10.7 Equally if datum 6 is 'popped' *from* the stack, as the last in and therefore the first out, the 'top' pointer will be reset to indicate datum 5. This, like so many things in computing, is lengthy to explain but such data movements and the amendment of the corresponding pointers happen very frequently and at very high speed without the user, in general, ever being aware of it.

10.10.8 If we have sent a number of jobs to the print-spooler we might wish them to be processed in the order in which they were prepared, by a first-in-first-out system, a very different operation of a stack.

When we represent this in similar fashion to our original stack, it is a little clumsy since, if we wish to push a datum into the stack we must first move the latter upwards in order to make space for the new datum to be added to the *bottom* of the stack. (The top and bottom pointers in Figure 10.5 are shown solid, for the present condition of the stack, and in outline for the changes which will take place to both if we push a new datum in or pop the first out.) Because of the awkwardness implied by this vertical sketch we often show a queue in a diagram as a *horizontal*, rather than a vertical, array but it is still commonly referred to as a **pushup stack** or **list**.

Figure 10.5 The clumsiness of a queue 'stack'.

10.11 DEQUES and wrap-arounds

10.11.1 For some purposes a queue may have data pushed and popped at both ends and you may find references to this as a DEQUE, or double-ended-queue.

10.11.2 Our final example of a stack is a **wrap-around stack** which may be seen as a kind of data merry-go-round in which a stack of fixed length is arranged so that the last datum cannot be pushed out of the end if the push-up or push-down goes too far but will, instead, transfer to the *first* datum position and so not be lost.

10.11.3 All such data structures are vital to the most fundamental operations of the machine in its *internal* handling of data and, in particular, for the storage and line-by-line running of our programs. Mostly these functions are transparent to us but job-scheduling, print spooling and many other operating system activities are intimately connected. They are, however, also extremely useful in applications programming. We can and should, especially if we have ambitions to be systems programmers, simulate the operation of stacks and queues by short program segments in high-level languages so that we may become familiar with all the principles involved.

10.12 Corruption by the machine

10.12.1 There are two very important little quirks of the machine which we must always remember. It does *not* 'clean up' after itself nor does it make sure that a location is empty before it pushes a data item into it! In practice when we pop or read a datum it is merely *copied* from the memory 'cells' in which it has been stored not (as is the case with a physical *parcel* in a stack), physically removed. We can only do this if we 'zeroize' the memory location, to 'clear out' the old item, after popping or reading the datum.

10.12.2 Even resetting the top-of-stack pointer leaves what *was* stored in the former top-of-stack location undisturbed, ready to corrupt future reads of that location if we are not careful. The exception is when, as in a push-up stack, each datum is moved into the position formerly occupied by a neighbour. Even then the 'last' cell will often contain its original occupant.

10.12.3 An equally serious problem is that when a datum is pushed or written into a memory cell it obliterates, beyond hope of recovery, what was there before. This is not a problem *unless we do it unintentionally and kill off essential data*. This is especially likely to happen with ill-designed sort routines.

10.12.4 As so often in computing another complication may arise when we are pushing or writing string variables. If we push a short string, say 'Invoice', into a location which previously held a longer string, say 'calendar' we *may* find, when we next retrieve the string, that it has become 'Invoicer'.

Can you see how? We need to check whether or not a string is automatically 'padded out' with spaces to overlay the whole of any previous string. Since it is usually *we* who define the individual string datum length to be allocated in any string array we must also be careful that we don't finish up with 'Invoi'.

10.13 Ordering and searching

10.13.1 The processes with which we are involved quite frequently are first the **ordering** of arrays, lists and stacks, and second, **searching** for individual data. (In both activities the danger of the corruption discussed in Section 10.12 is ever-present.)

10.13.2 Warning! One thing which we should never attempt to do, since each cell is directly related to all others in the same column and the same row, is to *order* a two- or more-dimensional matrix. Position, in a matrix, is *fixed* and is usually specified by the **labels** which we attach to rows and columns. Only *lists* of the various kinds may safely be ordered and even then we must, before doing so, very carefully check any mappings which there may be on to other lists! If two lists *are mapped on to each other* or linked in any such indirect way, extra care is needed to ensure that all necessary pointers are generated or that all moves of a datum in the one list are matched by corresponding moves in the other(s).

10.13.3 Ordering data arrays is usually described in computer jargon as 'sorting'. Strictly speaking this word really refers to 'classifying' items or putting them into categories, but the jargon is well established and we must use it. Be aware that it may cause some confusion.

10.13.4 Essentially all **sorts** are variations on putting the elements of an array into either *ascending* or *descending* order of either *magnitude* or *alphabetical order*. Since letters and all other symbols are held in the machine in numeric codes *all* sorts are, albeit transparent to the user, orderings by *numeric magnitude*. For *ascending* order sorts they are all variations on:

if $a_i \leq a_{i+1}$ then **leave** else **swap their positions**.

10.13.5 For *descending* order sorts we replace the $\leq$ sign by the $\geq$. We compare each pair of numbers in the array until the sort is accomplished. This is invariably done by finding **differences**. If $A - B < 0$ then $B > A$, and so on: all *ordering* relationships must be within the logical set $A > B$; $A < B$; $A = B$.

10.13.6 For string sorting we must insert an additional (nested) loop to make additional 'compares', letter-by-letter, if the two first letters are 'equal'.

10.13.7 In a simple 'ripple' sort we compare the first data item with each of the others in turn until we have the largest in a_n. Next we compare a_2 with each of the remaining cells, and so on. For an array of n items, with a total of $n - 1$ 'passes':

```
i = 1;
DO UNTIL i = n–1
j = i + 1
   DO UNTIL j = n
   If aᵢ > aⱼ DO SWAP
   j = j + 1
   ENDDO
   i = i + 1
ENDDO
END
```

10.13.9 The swap routine is a little complicated by the effect noted in 10.12.3, that pushing a datum into a location obliterates what was there before.

We see, in state **(i)** of Figure 10.6 the situation where n_1 and n_2 must be swapped in an ascending order sort. Moving n_1 into the cell occupied by n_2, as in frame **(ii)** of the diagram obliterates the '5' and leaves a '3' in both cells. Reversing the sequence, as in frame **(iii)** simply loses the '3'.

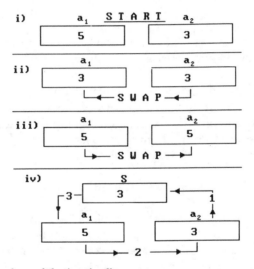

Figure 10.6   The place of the 'sort' cell.

10.13.10 What we must do is to declare a spare 'sort' cell, or cells if we are moving the contents of other lists at the same time, so that we can perform a 'rotary' swap, as in frame **(iv)**.

Here in move 1 the '5' moves to the 'sort' cell; in move 2 the '3' replaces the '5' as n_2 leaving the way clear for move 3, indicated by the dotted line, in which the '5' is moved from the sort cell into n_1 to complete the swap.

10.13.11 For short lists the ripple sort is very simple to program and relatively easy to understand. It does, however, need $n - 1$ passes and, on average, n swaps and there are many other sorts which are faster and more powerful. All have a strict mathematical foundation, as do the **searches**, by which we find a specific datum in an array. In order to search we must first know what we seek and then compare that 'key' with each item in a list until we find a perfect match.

10.13.12 In an n-dimensional matrix if $n > 1$ we can commonly search by a nested loop routine which compares our key with each cell in turn, row-by-column. This is because the data are unsorted.

10.14 The binary chop

10.14.1 In an ordered list, or linked lists, the process is much faster by a simple **binary chop** search where, on each pass, half the list is eliminated. Suppose that we have a list of 600 items to search. We start by testing our key against item number 300 and if this is larger than the key, we know that the 'top' 300 items can be eliminated from our search.

On pass number two we compare with the 'middle' item. No.150. Let us say that we find this to be smaller than the key – the lower half of the remaining list may be ignored.

Test item number 225 comes next and then either 262 or 188, as the case may be.

10.14.2 Can you continue the sequence if the key number is 288, for example? Why is this called a 'binary' chop? Can you write, and prove by desk-top running, a segment of pseudo-code which will perform this search? What is the minimum number of passes needed to find a single item in a list of 1000; 10,000; 1 million?

Once again we have examined the simplest such routine. There are many others which mathematicians have devised in order to search larger and larger lists, whether fully ordered or not. This is still the subject of research and new routines *are* still being developed but this happens comparatively rarely today since so many have already been discovered.

10.15 Trees

10.15.1 Quite often we have data in a list in a sequence which we need to preserve but which we need both to search quickly and to output in *ordered* form at will. A **data tree** structure lends itself to such management of data and in particular the **binary tree**.

We are all familiar with the typical growth pattern of many kinds of tree. From the root comes the trunk, or bole, and from the trunk come limbs from which in turn come branches which put out twigs from which grow twiglets and ultimately leaves or fruit.

10.15.2 In the much-simplified representation of a tree, in Figure 10.7, I have shown **two** limbs growing from the trunk, at level 2, **two** branches growing from each limb to form level 3, and so on.

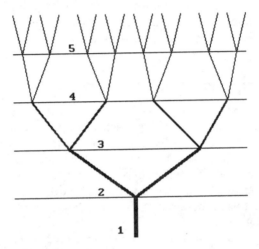

Figure 10.7   A simplified (binary) tree.

10.15.3 By restricting our model to two products of each 'split' we achieve two things. Firstly we can make each choice of path a yes/no decision, which fits with the binary logic of the computer.

Secondly we can bring into usable form a pattern of growth which is, in nature, too complicated and unpredictable to lend itself to clear modelling.

10.15.4 As always when we meet a new modelling technique there is some new language to master and here we find people in the industry speaking of a **forest** when our data structures include several trees, the **root**, or basic element of the tree from which all others grow, right up to the **leaves**, which may be the last **layer** or **generation** of the development of the tree.

10.15.5 The **branches** or **paths** are shown in Figure 10.8 along with the **nodes**. These are the key points in any such diagram and have been emphasized here.

10.15.6 At the *nodes* is where data are stored and where **branching** may take

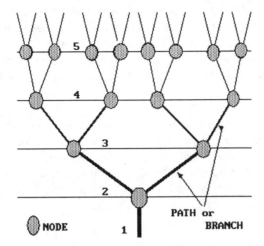

Figure 10.8   The tree labelled.

place in any of **two** directions. Only this last point determines a **binary tree** since in any other model, depending on its complexity, there may be more than two paths joining or separating at any node.

10.15.7 We are also familiar with the idea of a tree as a data structure in connection with our own families since a **family tree** is one way of making clear all the relationships through several generations. Such a tree, unlike what we have seen above, 'grows' downwards from the farthest traceable ancestor to the present youngest generation of children as shown in Figure 10.9.

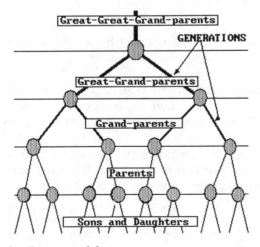

Figure 10.9   The family tree model.

10.15.8 All members of the family are **descendants** of the *root* from which they have sprung through several **generations**. In the example here I have gone to only five generations and kept to a binary model for the sake of simplicity.This is quite unlike most *real* families where multiple branching would be much more likely and the tree which results may be extremely complicated.

10.15.9 Nevertheless *this* is the model which is most commonly used to illustrate our use of trees, growing 'down' rather than up, with successive layers referred to as *descendants*. Many writers do, quite haphazardly, mix the botanical language of trees and the 'descendants' language of *family* trees but you may find it simpler to stick to just the one pattern until the subject is completely familiar to you.

10.16 'Growing' a data tree

10.16.1 Let us take a list of data and set about developing a tree from it. The most important point about the technique is that we use this special diagram (shown in Figure 10.10) at a node to hold a single datum and to give us the essential information to describe the tree completely in due course.

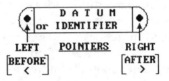

Figure 10.10 The fully annotated node.

10.16.2 The symbol is a rectangle which contains the datum and on each end of which is a semi-circle which is initially empty.

10.16.3 If a subsequent data item is less than, or comes earlier in sequence than, the node concerned, we indicate the necessary **branching (to the) left** by putting a small circle or a large dot into the left semi-circle of the node.

10.16.4 If a subsequent item comes *later*, or is larger, than the current datum we mark the right semi-circle in the same way. Since we are dealing exclusively with *binary branchings* this covers all the possibilities.

10.16.5 The circles or dots are like signposts pointing to left and/or right so we call them the **right** and **left pointers**. If either or both possible branches remain *unused* we leave the appropriate semi-circle(s) empty. If *both* semicircles are empty, then, we have reached a dead-end on that particular branching path.

10.16.6 We will follow the process with the data list:

> Malaysia; Namibia; Mauritius; Bahrain; Hong Kong;
> Pakistan; Sri Lanka; Botswana; Singapore and India.

10.16.7 When these are 'loaded' into our tree, in that order, the *root* will be Malaysia. You can see this in Figure 10.11.

Figure 10.11 The first descendant.

10.16.8 Item two, Namibia, comes *after* Malaysia in alphabetical order and will branch to the right, or, in the jargon, be a **right descendant** and we **set** the right pointer. **Note:** it is not always done in practice but I have *numbered* the nodes, in small circles alongside the datum, according to their places in the original sequence. This is only to help us keep track of what we are doing and is a good practice to follow until you are completely confident in developing trees.

10.16.9 Continuing down our list we come next to Mauritius, which is alphabetically before Namibia, but *after* Malaysia, hence it becomes a *left-descendant* of Namibia. The first three nodes, the root and its first two descendants, will look like Figure 10.12.

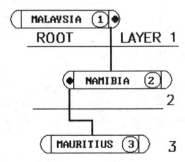

Figure 10.12 The first three nodes.

10.16.10 Bahrain is *ahead* of Malaysia in alphabetical order and will become the first **left descendant** of the former on layer 2, as shown in Figure 10.13.

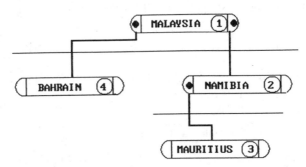

Figure 10.13   Developing the tree, ...

The diagram is beginning to get a little crowded and you will notice that the branch from node 1 to node 4 has been 'joggled' to save space. This does not affect the relationships and is often necessary.

10.16.11 Hong Kong is a *right-descendant* of Bahrain; Pakistan is a right-descendant of Namibia, whilst *nothing* in the remainder of the list comes 'before' Bahrain so *it* has no left-descendant. This process can now be continued until level 3 is completed as in Figure 10.14.

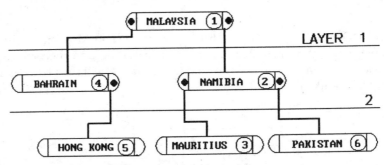

Figure 10.14   ... to complete layer 3.

10.16.12 Now it is a matter of following the same pattern of decisions until the tree is complete. Can you do so by identifying *and* placing the remaining nodes and setting the pointers to identify left- and right-descendants respectively?

10.16.13 When you deal with Botswana, which would appear to be a right-descendant of Bahrain since it comes *after* the latter, that position is already occupied, quite properly, by Hong Kong. You will need to 'slip down' a layer and attach it as a *left-descender* of the latter, which places it *between* Bahrain and Hong Kong, as it should be. Your finished result should resemble Figure 10.15.

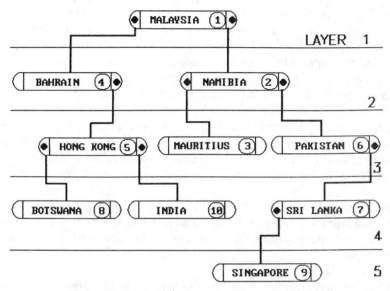

Figure 10.15   The tree diagram completed.

10.16.14 There are a number of important elements here. One is the idea of 'layers' or generations. We see that 10 data are contained in a maximum of 5 layers in the present case and this means that no search, to find whether or not the name of a given country appears in the list, will take more than 5 'passes'. Can you see why? Try testing to see whether Dubai is in the list; how many questions or 'compares' do you need? How many to check whether Singapore is listed?

10.16.15 The second important idea is that each pointer directs us to a numbered node, i.e. to a *specific datum*. From the root every datum is on an identifiable sequence of branchings.

10.16.16 A third important consideration is that when a datum has *no pointers* we need search no further; we have reached a natural 'terminator'.

10.17 Tables

10.17.1 Drawing a tree is an excellent way of organising our data in the planning stage, with pencil and paper on the desk, but it is not at all suitable for storage and manipulation in the machine as it stands. The first step is to put *all* the information in the tree into a table, as seen in Table 10.3, below.

10.17.2 There is one special problem which we need to consider. When a node has a pointer unused, i.e. when there is no left- and/or right-descendant, the pointer may be described as 'null'. To show this as a '0', which might be considered the natural thing to do, is misleading on two counts.

Firstly we *may* use a small circle, in the tree diagram, to indicate a pointer and using zero to indicate a null pointer would be confusing.

10.17.3 Secondly a number of languages allocate the subscript zero to the first row and column of an array since this makes the fullest use of available space. Since we may well refer to items in the table by their array subscripts this too could cause confusion if we had indicated null pointers by zero.

10.17.4 Since all nodes are indicated by positive integers and zero is *not* safely usable for the purpose it has become general practice to indicate a null pointer by −1. This may take a little getting used to but must be accepted as a practical rule.

10.17.5 Our tree, in table form, becomes as Table 10.3 and we may now consider how to refine this into an acceptable application of our normal data structures.

Table 10.3

Node	Datum	Left pointer	Right pointer
1	Malaysia	4	2
2	Namibia	3	6
3	Mauritius	−1	−1
4	Bahrain	−1	5
5	Hong Kong	8	10
6	Pakistan	−1	7
7	Sri Lanka	9	−1
8	Botswana	−1	−1
9	Singapore	−1	−1
10	India	−1	−1

10.17.6 At the system level, when used, for example, as a disc directory, the table, without the lines, can be stored as it is, so long as the length of each field is constant. This is the reason why many operating systems prescribe a *fixed length* file-name. Within a low-level language all symbols, whether alpha *or* numeric, are stored in numeric form and even the minus signs are inherent in the stored form as we saw in Chapter 6.

10.17.7 In most high-level languages however, since the data are **strings** and the pointers are **numeric** we have to seek other ways. The most obvious is to use an array since this already has a suitable rows and columns design.

However, we should need two **mapped arrays**, one for strings, another for numerics, when modelling in this form a tree such as ours.

We *do* gain a little, however, since the **row subscript** will automatically identify the **node** and the **column subscript** of the numeric array will tell us which are the left, and which the right, pointers. We can, therefore, both *save* one column and avoid the need for storing any labels to rows or columns.

10.17.8 As mapped arrays we shall have, for *our* tree:

$$
\begin{bmatrix}
\text{Malaysia} \\
\text{Namibia} \\
\text{Mauritius} \\
\text{Bahrain} \\
\text{Hong Kong} \\
\text{Pakistan} \\
\text{Sri Lanka} \\
\text{Botswana} \\
\text{Singapore} \\
\text{India}
\end{bmatrix}
\rightarrow
\begin{bmatrix}
4 & 2 \\
3 & 6 \\
-1 & -1 \\
-1 & 5 \\
8 & 10 \\
-1 & 7 \\
9 & -1 \\
-1 & -1 \\
-1 & -1 \\
-1 & -1
\end{bmatrix}
$$

in which all elements of our tree are reduced to the simplest possible model and are in a pair of *standard* data structures.

10.18 Back and trace pointers

10.18.1 We *can* go further and add **back pointers** to each node so that we may work *up* the tree from any node to trace its ancestors. As an example node 10 of our **countries** tree is descended from node 5 so that will be its *back pointer*. Node 6 has a back pointer of 2 since it is directly descended from that node.

10.18.2 We can also add a **trace pointer** which, from any node, will allow us to trace the next in alphabetical, (or numerical), sequence. Useful if, for example, we wish to list a directory in alphabetical order rather than in the alphabetically random sequence in which the files were created. In our example Bahrain points to Botswana, since that is the next datum in alphabetical order; hence Bahrain's *trace pointer* will be 8. (We can either treat the trace pointer of Sri Lanka, the last datum in alphabetical order, as *null* and make it −1, or as taking us back to the head of the alphabetical sequence to indicate Bahrain with the trace pointer set to 4, giving us a cyclic pattern which we could enter at any node. The former is more common but the latter is often useful.)

10.18.3 Here we see the arrays completed in this fashion; can you see how it all fits?

Malaysia	4	2	−1	3
Namibia	3	6	1	6
Mauritius	−1	−1	2	2
Bahrain	−1	5	1	8
Hong Kong	8	10	4	10
Pakistan	−1	7	2	9
Sri Lanka	9	−1	6	−1
Botswana	−1	−1	5	5
Singapore	−1	−1	7	7
India	−1	−1	5	1

10.18.4 Of the numeric array column 3 holds the back pointers and column 4 the trace.

As always familiarity will come only from experience and you should take a number of lists and try all the procedures for yourself.

10.18.5 Such a combination of arrays will not be operated on by any conventional matrix arithmetic, being merely data *holding* structures.

They are, in consequence, commonly referred to as **tables** and are of great importance at the system level. For example every disc has, as a fundamental part of its format, not only a directory but a File Allocation Table or Granule Allocation Table, showing how space is allocated to the file on disc.

There may also be a third part of the 'system information' written when the disc is formatted, the Hash Index Table, showing, from the hash code of the file name, exactly where the file is to be found in the directory. Even unusable tracks, detected on formatting, may be identified in a Track Lockout

Table. (I have shown the capital letters which form, in each case, the acronym most commonly used, for example the FAT.)

10.18.6 These are all tables, of just the sort we have been investigating, with file names and various pointers to indicate disc cylinders, tracks and sectors, unallocated disc space and the like. With the utilities available to us today to rescue, read or amend files on disc an understanding of tables and how they work is indispensable, even for getting the best out of a PC.

10.19 Networks

10.19.1 The final topic to be looked at in this chapter is another which becomes more important by the year, networks. Just as some types of data fit 'naturally' into tree structures so do networks have special characteristics.

10.19.2 Each is a structure which is marked by nodes and paths but a network is rarely so simple as to have a single root. The more common situation is to have several, or many, nodes with a variety of interconnections. As an example let us look at some simple computer networks.

10.19.3 Figure 10.16 shows three locations at which a company originally had computers installed on a stand-alone basis.

Figure 10.16 Four separate locations.

When it was realized that much of the software used, the data-flows, the information distribution and the file-handling were common to all users it was decided to install a simple datalink between head office and each of the subsidiaries, Figure 10.17. This worked well until, on several occasions, the system at *A* went down and *B*, *C* and *D* were left with the payroll not done on time.

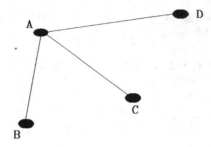

Figure 10.17   H.Q. to sub-stations.

The computer services manager then decided to duplicate some systems at *B* and to network *B*, *C* and *D* by additional links, shown in heavy lines in Figure 10.18.

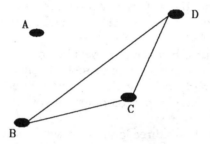

Figure 10.18   The 'safety-belt'.

This provided much needed security but was seen, after some time, to be rather slow and, since *A*, *C* and *D* were close together geographically, a high-speed LAN was set up to cover those three stations more effectively.

The new network is shown in Figure 10.19 (by means of curved lines to distinguish it from the others).

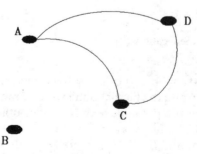

Figure 10.19   The LAN for *A*, *C* and *D*.

This too brought great improvements but to such an extent that the original slow link between *A* and *B* became an embarrassment and traffic around the other networks was too heavy to offer any relief to *B*.

By this time, however, technology had really leaped ahead and *B* and *A*, a considerable distance apart, were connected by a satellite link to extract maximum performance from all the systems as indicated in Figure 10.20.

Figure 10.20 The final link.

The final picture is shown in Figure 10.21.

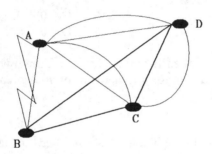

Figure 10.21 All the networks.

10.20 Networks as arrays

10.20.1 How could we model these different situations and evaluate the networks or cost out the solutions?

10.20.2 Once again the answer lies in using arrays of a rather special kind. The basic model is dominated by the fact that any single node may have a connection with any other node. What we develop is a **logical array** using a one to represent 'connected' and a zero for 'not-connected'. We can then model each network as a simple matrix.

10.20.3 For the first, star network, S =

	A	B	C	D
A	0	1	1	1
B	1	0	0	0
C	1	0	0	0
D	1	0	0	0

For the second, ring, (Figure 10.18), R =

	A	B	C	D
A	0	0	0	0
B	0	0	1	1
C	0	1	0	1
D	0	1	1	0

The LAN, (Fig. 10.19), gives us L =

	A	B	C	D
A	0	0	1	1
B	0	0	0	0
C	1	0	0	1
D	1	0	1	0

Lastly the satellite link gives L_1 =

	A	B	C	D
A	0	1	0	0
B	1	0	0	0
C	0	0	0	0
D	0	0	0	0

10.20.4 Look carefully at each of these, compare them with the diagrams and make sure that you can, for any network of your own devising, create the appropriate matrix.

10.20.5 For a combination of networks we may *add* the matrices. For example, $S + R$:

$$
\begin{bmatrix} 0 & 1 & 1 & 1 \\ 1 & 0 & 0 & 0 \\ 1 & 0 & 0 & 0 \\ 1 & 0 & 0 & 0 \end{bmatrix} + \begin{bmatrix} 0 & 0 & 0 & 0 \\ 0 & 0 & 1 & 1 \\ 0 & 1 & 0 & 1 \\ 0 & 1 & 1 & 0 \end{bmatrix} = \begin{array}{c} \\ A \\ B \\ C \\ D \end{array} \begin{array}{cccc} A & B & C & D \\ \begin{bmatrix} 0 & 1 & 1 & 1 \\ 1 & 0 & 0 & 1 \\ 1 & 1 & 0 & 1 \\ 1 & 1 & 1 & 0 \end{bmatrix} \end{array}
$$

10.20.6 In this case the matrix may be seen as wholly logical since each 1 indicates that there *is* a path from one node to another. It is, however, also an *arithmetic* matrix since the 1 also says that there is *only one* such path. What happens if we now also combine network L?

On the basis of the *arithmetic* we now see that there are, for example, *two* paths from *A* to *C*.

10.20.7 We can, in fact, use network matrices *either as logical or as arithmetic* tools so long as we don't confuse the two very different types of process.

10.20.8 Can you complete the $S + R + L$ matrix as well as all those representing any combination of two or more of our example networks? What is the array which represents *all* the networks together, as in Fig. 10. 21?

10.20.9 As a final exploration of the ideas examined in this chapter discuss with your colleagues how to exploit these models. Invent and model some networks of your own. Prepare simulations of the traffic on each network, inserting numbers of your own choosing to identify the 'data-packets-per-day' flow rates. Allocate a price per packet for each path on each network and *cost* the traffic. Identify any intrinsically weak or possible redundant links in the system. All of these things may be done using the arithmetic of matrices which we covered in Chapter 9.

10.21 Conclusion

Throughout this chapter we have seen how common *existing* mathematical techniques provide ways and means of creating and manipulating data structures. We have also seen the need to practise working with them to achieve the understanding and fluency of their use which will allow us, from the earliest design stage of programs, to *select* existing and reliable ways of doing things rather than having to 're-invent the wheel'.

Exercises

(1) A manufacturer produces three different machines, A, B and C each of which embodies different arrangements of the sub-assemblies, 'd', 'e', and 'f'. Each of these, in turn, is made up of different quantities of parts having the part numbers 101A, 101B and 101C and which cost $3, $2 and $4, respectively. Model A consists of 3 of sub-assembly 'd' and 3 of sub-assembly 'e'. Model B consists of 2 of 'd' and 4 of 'e'. Model C has 3 of 'd' and 5 of 'e'. Of the sub-assemblies: a 'd' needs 3 of 101A, 4 of 101B and 3 of 101C an 'e' needs 2 of 101A, 3 of 101B and 5 of 101C.

(a) Put all these data into appropriate matrices.

(b) What is the matrix of costs/machine models?

(c) (i) In one day 80 of A, 50 of B and 40 of C can be produced. Set these data into the appropriate vector.

(ii) Find, by matrix operations, the total value of one day's production.

(iii) If the costs of the parts 101A, 101A and 101C were increased by 5%, $7\frac{1}{2}$% and 4%, respectively, what would be the total increase on one day's production costs?

(2) Why are linked lists often used instead of arrays?

(3) Why are such lists linked?

(4)   (i) What is the function of a pointer?

(ii) Show, in a diagram, a linked list of your own devising complete with pointers.

(5) In what ways do LIFO and FIFO stacks differ?

(6) (i) Identify the characteristics of one kind of sort.

(ii) Illustrate your chosen example by a suitable diagram.

(7) Illustrate the process by which a binary chop search would find item 77 in an ordered file of 150 records.

(8) (a) Construct a data tree to hold the data: 497, 235, 502, 50, 79, 143, 64, 379.

(b) Create, for the same tree, a table with the appropriate right- and left-pointers.

(c) Add the necessary back- and trace-pointers.

(d) Show how the table in part c) would look after the insertion of 128.

(9) The following matrices, A and B, show communication links between three separate locations

$$A = \begin{bmatrix} 1 & 0 & 0 \\ 1 & 0 & 1 \\ 0 & 2 & 0 \end{bmatrix}, \quad B = \begin{bmatrix} 0 & 1 & 0 \\ 1 & 0 & 1 \\ 1 & 0 & 1 \end{bmatrix}$$

(a) Draw the corresponding networks.
(b) Add the matrices and draw the resulting network.
(c) Which, if any, of the communication links are one-way only?

(10)

$$\begin{bmatrix} 0 & 1 & 0 & 1 \\ 0 & 0 & 1 & 1 \\ 0 & 0 & 0 & 1 \\ 0 & 1 & 0 & 0 \end{bmatrix}$$

is the network which at present connects various workstations to a central file-server.

(a) Label the matrix and show which is the central computer.
(b) It is planned that the three 'outstations' shall have full local processing capability but shall become fully linked to each other by two-way communications whilst remaining connected to the central file-server.

   (i) What must the new matrix be?
   (ii) Write down the matrix which shows what new paths are needed.

11

Numerical and algebraic methods

Objectives

After working through this chapter you should be able to:

❑ use elementary numerical and algebraic methods to model and simplify
   problem solving
❑ understand iterative, recursive and convergent processes.

11.1 The background

11.1.1 I have said earlier in this book that the mathematicians of old dream-
ed of having a calculating device of real power. Among their number were
people such as Babbage in England, Pascal in France, Leibnitz in Germany
and, more recently, Hollerith in the United States.

11.1.2 In the absence, until very recently, of any truly successful device
much important work in modelling the real world and solving real problems
was both laborious and time-consuming.

11.1.3 Mistakes in calculation were also a major difficulty since with paper
and pencil, or even small mechanical calculators, the human being was, and
still is, relatively unreliable if called on to spend large periods of time concen-
trating on accurate computation.

11.1.4 Because of this arithmetic was traditionally considered a low-grade
task and the ultimate mathematical skill was to reduce a problem to an

algebraic model, or a Euclidian *geometric model* and to operate on these models in ways which were essentially *abstract*.

11.1.5 These were rarely of use to anyone with a *practical* task to perform, like the builder or the engineer and there were many such models which could not be fully 'solved' or proved, or which could not be reduced to a suitable form for real-life applications.

11.1.6 There were also many problems to which no *exact solution* could be found but 'final' solutions, full proofs or exact solutions are not necessarily what any builder, engineer or accountant needs. We need, for most practical applications, *working values* which allow the persons concerned to get on with solving their real-life problems.

11.1.7 It is in meeting this need, in *real* applications of mathematics, that the roots of the development of **numerical methods** lie. Today, in many applications of the digital computer, they come into their full usefulness.

11.2 Language

11.2.1 As always there are some special words and symbols associated with the topic and among these are **iterative, recursive, convergence** and **limit** as well as, from Chapter 5, **error** and **error bound**.

11.2.2 We will deal with each of them more fully as we meet them in action but 'iterative', one of the most important to us, really only means 'repetitive'. An 'iteration' is a process which repeats. Often it repeats the same core calculations to arrive at ever-closer approximations to the value being sought. *Many computer programs are iterative.*

11.2.3 **Approximation** certainly does not mean a wild guess, or even an estimate and we should not fall into the common trap of assuming that an approximation is an inferior solution to a problem. It is often the only useful or usable solution which exists! For π, for example, we can never have a final value, only an approximation, to an absolute error bound which is close enough to satisfy the practical needs of the user.

11.2.4 Every rounding, truncation or 'significant figures' answer results in an approximation. *The key is the level of accuracy which is of practical value to the user.* For a great many calculations of all kinds we define the appropriate degree of refinement of an approximation by defining the acceptable error bound. A size calculated 'correct to seven decimal places, as in seven figure tables, for example, may be over-elaborate for an engineer whose instruments or machines have an inherent error bound of ±0.001 mm. He may, however, need such precision *within* calculations in order to minimize error as we saw in Chapter 6.

11.2.5 Recursive, for our present purpose, means only that the result of any step of an iteration is 'fed back in' to the calculation for the succeeding step. When successive values from an iteration show *decreasing differences* we say that the iteration is 'converging'. It is not difficult then to interpret '*divergence*'.

11.2.6 We met the word 'limit' earlier when it described the theoretical end of a process, such as finding the value for π, even though we may never be able to arrive at a finite answer.

11.2.7 We may also *limit* the number of repetitions of an iteration by defining the acceptable error and from this the degree of accuracy to which it should pursue its calculation.

11.2.8 The principle new *symbol* which we shall use is Δ, from the Greek letter delta, representing the *difference* between any two numbers or successive terms in a sequence.

11.3 Iteration

11.3.1 Let us examine a simple iterative process and see what happens in a pencil-and-paper 'run' by tabulating the results of each 'pass' round the loop. These are the vital techniques which should be followed when exploring any iteration (and represent a 'standard' program test routine):

```
i = 1, d = 1, s₁ = 1
Read n
   DOWHILE i ≤ n
     Print i, sᵢ
     i ← i + 1
     d ← d + 2
     sᵢ ← sᵢ + d
   ENDDO
END
```

11.3.2 A table of results is given in Table 11.1

11.3.3 This table contains some repetitions since *i* is repeated in the print-out but it is the *pattern* of table which we should follow. It will, in many cases,

Table 11.1

PASS$_i$	d	Print i	Print s_i
1	1	1	1
2	3	2	4
3	5	3	9
4	7	4	16
5	9	5	25
6	11	6	36

be necessary to do more than six 'sample' passes in a run but six is enough for us here.

What does the 'print' information tell us?

What function maps the i onto the s_i?

11.3.4 Clearly it is the table of natural squares which we are developing so how did I know that the model in the pseudo-code would generate that function? Let us take only the 'print' columns and add a differences (Δ) column so that we may fully explore the iteration. (The table of any such sequence should, ideally, start at zero since this is the origin of all our natural numbers.) The table including differences is given in Table 11.2 and we see that the sequence of differences is the set of odd numbers. In other words the natural squares of numbers up to n may be found by adding all the *odd* numbers from the 1st to the nth. It is not difficult to see how the table would continue.

Table 11.2

i	0	1	2	3	4	5	6
s_i	0	1	4	9	16	25	36
Δ	0	1	3	5	7	9	11

11.3.5 There are many ways of modelling this situation and one of the traditional mathematical models is some variant of:

$$n^2 = \frac{n}{2}*\left(1+(2n+1)\right)$$

11.3.6 This is a fair comparison between 'classical' mathematical solutions and the straightforward, direct, iterative model in our pseudo-code segment above. Such a model is usually both the simpler and the more efficient way of using the power of the computer and does not involve memorizing formulae.

11.3.7 Note that in this case we are dealing with the natural squares, which are all integers, so there is no question of approximations in our iteration.

11.4 An iterative model

11.4.1 There is not a set of mathematical tables stored in the computer to provide square roots, for example, even if the programming language offers functions such as 'sqrt(7)'. How could we model a process for finding $\sqrt{7}$? Let us try a few 'what if' steps and see what happens.

11.4.2 What if we take a first estimate, e_1, of $\sqrt{7}$ as being 3.5? If we now divide 7 by e_1 the answer will identify another factor of 7. If the *second factor* were to turn out also to be 3.5 then that would be $\sqrt{7}$. Instead it is 2.

11.4.3 We can now see that 3.5 is a *larger* factor of 7 than is $\sqrt{7}$ and 2 is *smaller* so $\sqrt{7}$ must lie somewhere *between* 2 and 3.5. We can try the simple *mean* of these two values, 2.75 as a closer approximation. Now let us turn to the calculator.

11.4.4 If we now divide 7 by our improved estimate, 2.75 we get 2.5454545, and again we know that the true root must lie between these two values. Once more we 'average' these, giving 2.6477272 and 7 divided by this gives 2.6457520. Averaging these two gives 2.6457506; the next repeat gives 2.6457513 and dividing 7 by this gives 2.6457513. These are equal so that we have, in a very few steps, with a calculator giving seven places of decimals, found $\sqrt{7}$ correct to 6 decimal places.

11.4.5 If e_i is each successive estimate and n is the number whose root we seek, then $e_1 = n/2$ and, putting the steps above into a formal model:

$$e_{i+1} = \frac{\left(e_i + (n/e_i)\right)}{2}$$

we have the core of our program segment.

11.4.6 Test this model carefully, using any, or all, of the steps detailed above, and with 7 as n.

11.4.7 Now to tabulate those steps, *with* the differences column as shown in Table 11.3.

11.4.8 When we have such data it is often helpful to plot a graph, in this case using the successive values of the estimated root and it's second factor

Table 11.3

$i =$	e_i	n/e_i	e_{i+1}	$e_i \Delta e_{i+1}$
1	3.5	2	2.75	0.75
2	2.75	2.5454545	2.6477272	0.1022728
3	2.6477272	2.6437768	2.6457520	0.0019752
4	2.6457520	2.6457506	2.6457513	0.0000007
5	2.6457513	2.6457513	2.6457513	0.0000000
6				

of 7 (see Figure 11.1). From this graph we can see the rapid convergence of such an iteration as we might expect from the rapid decrease in the differences, in the fifth column of Table 11.3.

Figure 11.1 The convergence is clear.

11.4.9 This is the typical behaviour of such a routine and it is often extremely valuable when programming to arrive, initially, at such a very simple model, sometimes referred to as a 'boiler-plate' version, which is simple, works effectively and is easy to test.

11.4.10 The model is fast, even if our initial estimate is, as in our example, very crude. An initial estimate, sometimes referred to as a 'seed', *must* be provided by some means and many textbooks recommend putting in an estimate 'by hand', so to speak. It is, however, extremely easy to halve our n in order to arrive at a first estimate and it is often the case that a very simple method is good enough.

11.4.11 Once we have established the working core of our program we can, if need be, refine our model or add other steps to it (for example to accept *any* n). We may even seek out a more elegant or more efficient version from classical mathematics texts. Any specialist textbook on numerical methods will contain a wealth of such models, so long as we know just what we are looking for.

11.5 Limiting the iteration

11.5.1 We need to add a 'stop' to our procedure, and also include some tests, as it is all too common for a computer to 'lock up' when performing an iteration which has no natural end or which is, in fact, *diverging*. This will often lead to an 'infinite loop', a long-standing hazard in programming!

11.5.2 The hazard of divergence is best dealt with by comparing successive *differences*. As we see clearly in our *convergent* example the differences *always* decrease. If, on any pass, the difference has *increased* the iteration is diverging and if this is not what is expected we must have a subroutine which will cause our program to break out of the loop *and let us know what is happening*.

11.5.3 The second need, to stop the procedure when it has gone far enough for our purpose, means that we must first define, with the user, what that purpose is and *to what level of accuracy of estimate* the answer is to be calculated.

11.5.4 If we need an answer correct to three decimal places what will be the absolute error bound? If you are in doubt refer back to Chapter 5.

In order to be sure that we are rounding correctly we would ordinarily calculate to four decimal places (but, for some purposes, we may wish to calculate to as much as, but certainly no more than, five decimal places.)

11.5.5 We achieve this by controlling our loop with a statement such as:

DOWHILE abs $(e_i - e_{i+1}) > 0.0001$.

where the 'abs' (absolute) function gives the difference between two values regardless of whether the difference is positive or negative.

11.5.6 You may now feel that you could write a pseudo-code routine for 'sqrt(n)' which embodies a test for divergence and the 'stop' when an answer can be given correct to three decimal places.

Try it; vary the n and test each version first by a desk-top run using your pocket calculator and then as a short program in your chosen high-level language on a computer. As always mathematicians have devised faster methods for finding roots and further study will lead you to these if you are interested.

11.6 Aberrations

11.6.1 The problem of controlling an iteration should not be underestimated and much of modern 'chaos theory' has evolved from the study of iterative routines which produced unexpected results. Often such aberrations show up as an unexpected failure to converge or a lapse, at an unexpected point, into results of *disorderliness* where order was expected.

11.6.2 We must, therefore, be meticulous in auditing the operation of an iterative process. We can do this by printing full listings of intermediate results if necessary; by graphing sample sets of results and looking for anomalies in the graph; by continuing the iterations longer than may seem strictly neces- sary; by paying careful attention to tables of differences or by trying 'unorthodox' seed values or step sizes.

11.6.3 *Test data should always include unlikely or even impossible values to make sure that they are detected by our programs before generating misleading reports.*

11.7 Simultaneous equations

11.7.1 Let us look at a rather more complex type of problem and see what we can do with it.

11.7.2 Here is a pair of simultaneous equations:

$$\text{(a)} \quad 2x + 3y = 5$$
$$\text{(b)} \quad x^2 + y^2 = 4$$

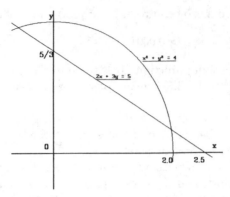

Figure 11.2   A pair of simultaneous equations.

for which we are to find *positive* values of x and y which satisfy both. (Since we want only positive values I have shown only the first quadrant of the graph.)

11.7.3 You may find it useful to draw the graph to a large scale for reference as we work through the example.

11.7.4 This is exactly the kind of problem which presents quite difficult algebra but yields to an heuristic, iterative, numerical approach. The (+,+) point at which the straight line cuts the circle is one at which both equations are true.

11.7.5 At that point x is clearly a little less than 2. If we take a 'seed' value of 1.5 and substitute this for x in equation (a) we shall find that $y = 0.66667$ to five decimal places. Substituting this for y in equation (b) we find that $x = 1.88561$ and we take this as x_2 for our second pass and substitute it for x in equation (a) to find y_2, which, in turn, we substitute in equation (b) to find x_3, and so on. Note that, in all these cases, we must be careful not to confuse the *subscripted* values of x or y with *powers* of those values.

11.7.6 In Table 11.4 below I have set out the sequence of values at each pass and you will see that, even for a computation of admitted algebraic complexity the numerical method leads us to a result very quickly indeed, and we see that $y = 0.3544$ and $x = 1.9683$ satisfy both equations.

11.7.7 You should now try to write a program segment to include this routine based on a simple loop construct. The model which we have been looking at is clearly both recursive and convergent and this should guide you in your programming.

Table 11.4

i	Equation (a)		Equation (b)	
	x	y	y^2	x
1	1.5	0.66667	0.44449	1.88561
2	1.88561	0.40960	0.16777	1.95761
3	1.95761	0.54239	0.29419	1.92505
4	1.92505	0.38330	0.14692	1.96293
5	1.96293	0.35804	0.12820	1.96769
6	1.96769	0.35487	0.12594	1.96826
7	1.96826	0.35449	0.12566	1.96833
8	1.96833	0.35445	0.12563	1.96834
9	1.96834	0.35444	0.12563	1.96834

11.7.8 Now try putting into practice an alternative version of the model, we could call it 'routine 2', in which you put the seed value into equation **(b)** and transfer the result to equation **(a)** each time. What do you notice about the results?

If you arrive at $x = -0.4299$ and $y = 1.9532$ to four decimal places we agree.

11.7.9 Look again at the graph; can you see what these values might correspond to? They are the x and y values which satisfy the two equations at the other intersection.

11.7.10 However we *specified* at the outset that only the positive values were applicable so that if we had tried only this second method we would have defective answers.

11.7.11 We could avoid this by including in our model a standard test couched in terms something like:

if $x \le 0$ OR $y \le 0$ then CONTINUE else DO routine 2.

As mentioned earlier the auditing of our model is very important and will often cause us to add several conditional tests to a simple routine.

11.8 Expansions

11.8.1 Many functions are **expansions** which seem a little different at first but which are, again, not too difficult to program. For example a very useful approximation to π is to be found by the model:

$$\pi \approx 4 * \left(\frac{1}{1} - \frac{1}{3} + \frac{1}{5} - \frac{1}{7} + \frac{1}{9} - \frac{1}{11} \ldots \right)$$

which is really quite easy to write, since we can predict further terms without difficulty, but rather tedious to work by pencil-and-paper arithmetic.

11.8.2 Left to itself such an iteration will recur indefinitely and so we need, once again, to consider the level of precision which our application needs in order to set a 'break-out' difference between successive terms which will stop the iteration when it has gone far enough. If you try to model this you may be surprised at the ease of calculating π although the iteration is quite slow compared with our $\sqrt{n}$ model, above. However, *absolute* speed does not really concern us since a thousand passes round such a loop (and all iterations involve the loop construct) take little perceptible time when it is the computer which is doing the 'number-crunching'.

11.8.3 Virtually all the angle functions, logarithms and many other mathematical 'tables' are calculated using such iterative expansions, although, as with π, mathematicians have provided us with faster and more powerful models than some of the relatively simple ones at which we have looked in this chapter.

11.9 Curved areas

11.9.1 The last topic in this section of our work deals briefly with some questions concerning curved-line graphs. There are two questions which arise frequently when we need to deal with functions of the kind which we saw, in Chapter 7, to generate curves when plotted on a graph.

11.9.2 The first such question is 'how may we **interpolate** or **extrapolate** from what we know?'.

11.9.3 The second question is, 'what is the area of a surface, like a graph, which has one curved edge?'.

11.9.4 In the past both such questions were dealt with by rather elaborate methods but we may, for most purposes, use our iterative processes to good effect.

11.9.5 If we take, as a basis for experiment, the graph of $y = ax^2 + c$ we can write a standard program module to calculate the curve:

```
Read x, breakout, step, a, c
   DOUNTIL x = breakout
      y₁ ← (a * (x ↑ 2)) + c
      print x, y₁
      x ← x + step
      y₀ ← y₁
   ENDDO
END
```

We can now calculate the points of our graph by initializing the variable x to our first value, 'breakout' to the last required value and a and c, which are constants for any given function, to their fixed values.

11.9.6 Furthermore we may make our curve as smooth as we need by using a very small 'step' value, that is the amount we add to our previous value of x to derive the next value.

11.9.7 As a little 'bonus', if, as we stated earlier, we wish to find the **area** under the curve between any two values of x the step size will also be the *width* to be used in our calculation. Let us look at the diagram of a small section of the area under a curve in Figure 11.3.

11.9.8 To find the area, A, under the curve between any two x values, using 'firstvalue' and 'secondvalue' as the variable names for the two values of x, for the moment, we can insert into our program segment a subroutine of the type:

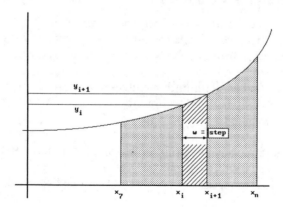

Figure 11.3 The area as strips.

DOWHILE x ≥ firstvalue AND x ≤ second value
    A ← A + step*½$(y_i + y_{i-1})$
ENDDO

11.9.9 We can do this because, as we see from the diagram, the area between, say x_7 and x_n (lightly shaded in Figure 11.3), may be seen as the sum of a number of strips, of which a specimen is shown with diagonal hatching in the diagram.

11.9.10 Each strip is a close approximation of a **trapezium** for which the **area** is the width multiplied by the *average* of the two parallel sides.

11.9.11 In our example our main routine gives us sequential pairs of the *y* values, the heights, and the step size gives us the width. Since all three occur in each pass of our original loop it is only a matter of 'capturing' existing data and accumulating partial areas under *A*.

11.10 Interpolation

11.10.1 If we wish to *interpolate* further values between values for *y* which we have already plotted, or *extrapolate* from them we need only set up our program segment so that we may *enter specific values of x* from the keyboard.

11.10.2 As always it is practice with such routines which will bring confidence and above all don't be afraid to try out variations, such as finding *x* from an input value for *y*.

11.11 Some algebraic methods

11.11.1 We have found ourselves increasingly using *symbols*, rather than numbers, and constructing *complete* rather than partial models. It seems that we have reached a natural point at which to look at some **algebraic methods**.

11.11.2 Let us first of all remind ourselves that the main purpose of algebra for us, as for most *users* of its techniques, is to allow us to create **complete models**, which we can then program.

11.11.3 Such models will include all *known* information whether **constants** or **variables**, which will eventually allow us to isolate and determine the *unknown(s)* which represent the usable solution to a problem.

11.11.4 In computing it is usually worth developing such a model only if it represents a problem which arises frequently, such as calculating employees' pay. Not just *one* person's pay but any number of people's on any number of occasions. It is *this* which leads us to algebra and its methods. If my own pay were $10 per hour and I had worked 40 hours with $15 taxes to be deducted the *arithmetic* model, $(40 * 10) – 15 would be sufficient.

11.11.5 In order to **generalise** the model, that is to say, make it useful for *any* worker in *any* week for *any* tax deductions we would have to describe a more general model such as:

> *R*ate times *H*ours minus *T*axes is the take-home *P*ay.

This is a true representation of the situation but still rather cumbersome and easy to misread so we reduce it to *abstract symbolic form*, and say $R * H - T = P$.

This is a true **algebraic model**; a **full** statement of the model in **symbolic form**.

11.12 What the model means

11.12.1 In one cell of a spreadsheet which I use frequently appears a 'formula', as it is described in the language of spreadsheets, which seems quite complex:

@INT(B1 + 0.5 + (F1 – A1) * (D1/C1)).

We should not ordinarily recognize this as an equation since it lacks the very symbol which says 'equation', namely the '=' sign to denote that the quantity to its left has precisely the same value as that to its right. (We have seen earlier that there are also many *in*equations', or 'inequalities' denoted by $<, \leq, >, \leq$, $\neq$ and also many approximations denoted by $\approx$ (or $\simeq \doteqdot$ perhaps, depending on the country) and we must by no means forget these).

11.12.2 In our spreadsheet example the equivalence is **implied**, that is to say that, although it is not *stated*, the cell in which the formula is placed is, effectively, the left-hand side of the equation and the destination for the value to be calculated.

11.12.3 What all this amounts to is that there is a specially adapted *spreadsheet algebra*.

11.12.4 Let us look at it in detail and find out what all the parts of

@INT(B1 + 0.5 + (F1 – A1) * (D1/C1))

actually mean.

11.12.5 First we can identify the @ symbol as being one of the ways in which spreadsheet software identifies a *calculation which is to be performed* with the result being placed in the cell where the formula is located.

11.12.6 INT is a common computer-language command to take the **integer part** only of the answer by *truncating* at the decimal point. Hence INT(3.845618) would return the value '3'.

11.12.7 The rest of the model is in brackets and so must be dealt with by the 'rules of brackets and the sequence of operations'. These state that the contents of brackets are to be dealt with in sequence from the *innermost to the outermost* and that arithmetic operations are to be done in the sequence:

(1)   Raise to power, (x^2), or $\sqrt{}$
(2)   * or /
(3)   + or −

and when two operators are of equal 'precedence', say 12 * 4/3 they are to be performed in sequence from **left** to **right**.

11.12.8 If we are in any doubt about precedence, especially when designing a program model, the remedy is to break the model into single operations within yet more pairs of brackets. Even many pocket calculators today will allow us to 'nest' brackets six 'deep'. One precaution is to make a final check to ensure that the number of left-brackets, '(', exactly equals the number of right-brackets, ')'. Many seemingly mysterious error reports arise in computing because of a failure to check this.

11.12.9 We can label the model to indicate the steps in which we first solve (F1 − A1); next (D1/C1); then what is inside the outermost pair of brackets, shown with the 'long' underline until *finally* we can take the **integer** value of the result. This is shown in Figure 11.4.

11.12.10 There remains the mystery of the F1, A1, D1 and so on. In the algebra of spreadsheets these are the **cell addresses** in which the variables are stored and to this extent they are precisely similar to the storage of *declared*

$$\text{INT} \, \dfrac{\left(B1 + 0.5 + \overset{\text{1st}}{\overbrace{(F1 - A1)}} * \overset{\text{2nd}}{\overbrace{(D1/C1)}} \right)}{\underset{\text{3rd}}{}}$$

Figure 11.4   The sequence for solving the model.

variables or constants at fixed addresses in the computer's memory. They are, then, just **variable names** rather than the x, y and z of conventional, wholly abstract, algebra.

11.12.11 We can now see that (F1 − A1) says 'take the value stored in A1 and subtract it from the value stored in F1'. What is *not* said is that the result will have to be put into *temporary storage* until it can be used again later in the calculation. Such temporary storage, exactly like writing intermediate values on a piece of paper in everyday practice, is often referred to as **scratchpad** memory and many applications of our own programs need some form of scratchpad, or temporary, 'holding' variables, built into our planning.

11.12.12 In the spreadsheet all the manipulations of such models are transparent to us but it is *we who have to design the models.*

11.12.13 It is exactly here that we need to have some understanding of algebraic methods; they are, quite simply, the language of modelling for the computer. The original of the spreadsheet operation which we have been considering is the equation:

$$x_i' = \overline{X}' + \left(x_i - \overline{X} * (\sigma' / \sigma) \right)$$

11.12.14 We may now be able to see how this original 'maps onto' the spreadsheet version. Discuss this with your colleagues, try some values for all the xs and σs and see also if you can work out what the 0.5 is for in the spreadsheet version.

11.13 Notation

11.13.1 When looking at any symbolic model the most important thing is not to be put off by the symbols themselves. They all merely *stand for* something and are not sacred objects! What we need is patiently to unravel these labels and, if they are unfamiliar to us, to look them up in a reference book.

The symbol σ, for example, is commonly used in statistics for standard deviation (SD), but we could, for our own purposes, use any symbol we choose. The great advantage of using σ is that it is understood anywhere in the world as SD when it is used in a statistical context.

11.13.2 In a private model or program we could perfectly well call it 'Fred' if we wished! What we **must** do, for the sake of *maintainability* of our programs, is to include our chosen symbol, the word or words which it stands for, the type of variable used and any other relevant information in our **data dictionary**.

Even we, looking at a program which we wrote a few months earlier, may find it difficult to interpret unusual symbols or 'Freds' if we did not document them properly.

11.13.3 In traditional mathematics the documentation takes the form: 'let x = the rate of pay; let y = the number of hours and let z = the amount of tax'.

11.13.4 The use of x, y and z is a tradition in formal algebra largely because the early letters of the Arabic alphabet were already being used for arithmetic or geometric 'identities' such as A for area, B for breadth, C for circumference, and so on.

Since letters up to t for time were already in use for *known* values mathematicians adopted the practice of using x, y and z for the *unknowns*. For several centuries, three-variable relationships were the most complex to be tackled and these three symbols served, and continue to serve, well enough. You must not let these old habits worry you if algebra is not one of your favourite pastimes; any symbol which is appropriate to the task will do equally well.

11.13.5 We have only two other basic ideas of notation to discuss at this stage and the first is the convention for multiplication. There are *two* common ways of indicating multiplication; $a \times b$ and $a * b$ but we have already seen that the first, older version, is of no use in computing. This is for two reasons. Firstly we have no means of distinguishing between the *quantity* 'x' and the *operator* '$\times$' and secondly because the computer keyboard and the translator software cannot recognize the difference in *appearance* of the two ways of writing the letter 'x'. In mathematics if something *can* be misunderstood it is treated as *wrong* since, sooner or later, it *will* be misread or misinterpreted.

When x came into use in algebra a similar problem arose and multiplication was indicated by $x.y$, with the dot on the line indicating 'multiply' and this still applies in mathematics today. Little by little, especially with the invention of decimal fractions, it became clear that even the dot could be ambiguous and it was not safe to write 3.5, for example, to mean 3 *times* 5.

11.13.6 The solution adopted was to establish that where different symbols represented quantities to be multiplied they could be written side-by-side without *any* intervening operator. Nowadays it is common practice to write $x * y$ simply as xy, but *not all translator software recognizes this* so be cautious!

11.13.7 In similar fashion $x(y + z)$ means that *all the contents* of the brackets are to be multiplied by x either before enumeration, as $xy + xz$, or, if we know the values to 'substitute' for x, y and z, say 2, 7 and 5, respectively, $2(7 + 5)$ we may work as $2 * 7 + 2 * 5$ *or* $2(12)$, whichever is the more convenient.

11.14. Powers of numbers

11.14.1 We then have a very confused situation if we wish to multiply a number by itself, for n times, and we try to write it as, say, xxx. Do we have $x * x$ or $x * x * x$? If x were 5, for example, the answer could be 25 or 125 depending upon our interpretation. As we have seen before we firstly need to distinguish between '×' meaning multiply and the (unknown) quantity 'x'.

11.14.2 Next we borrow from the very old ideas of 'powers' of numbers in our column system (and in square numbers and their relatives) and use x^2 to denote $x * x$, $3x$ to denote $x + x + x$ and x^3 for the $x * x * x$.

11.14.3 This **index notation**, or 'powers of n' is very common in algebraic models and leads to a number of special laws governing the use of 'indices', which is the plural of 'index'. In the notation x^n the small numeral in the 'top-right-corner' of the term 'x' is its **index** and indicate how many xs are to be multiplied together.

11.14.4 The first new law is that 'when like terms in index notation are multiplied together their indices are *added*'. For example $x^2 * x^3$ is equivalent to $x * x * x * x * x$ or $x* x * x * x * x$ which is x^5.

11.14.5 Here I have mentioned, for the first time, the phrase 'like terms', which is important in algebra.

In, for example, $2x^3 + x^2 + x + 5x^3 + 3x + 4x^2$, we see that all the terms are similar since they all include x. However x^2 is not exactly comparable with x^3 since it is on a different scale.

11.14.6 To simplify the original pattern we can, however, rearrange the order so that we have $2x^3 + 5x^3 + x^2 + 4x^2 + x + 3x$. We call this 'gathering like terms' which we do, both by tradition and for clarity, in the ordered pattern of alphabetical order, e.g. x before y, and within that, descending order of indices.

11.14.7 It is not too difficult to see that we now have, grouping the terms altogether, $7x^3 + 5x^2 + 4x$, which is rather simpler than our original version.

11.14.8 Such **simplifying** is at the very heart of algebraic methods. When we have a number of terms in which different symbols are multiplied together, for example $3abc + 2ab + 5cba$ it may be a little more difficult to decide which to 'gather'.

Only the '*abc*' and the '*cba*' (or any *other* arrangement of these three letters under multiplication) are 'like' since $c * b * a$ comes under the **associative law** and could equally well be written as $c * a * b$, or $c * b * a$.

Our gathering of like terms, then, will result in $8abc + 2ab$ but, by the **distributive law**, since we have, in effect $(ab * 8c) + (2 * ab)$, we *could* write this as $ab(8c + 2)$.

11.14.9 Think about this, discuss it with your colleagues, try a variety of such associations and distributions until you are confident with them. They are all ways in which we can simplify our models. The laws of arithmetic operate just as powerfully and to great advantage in algebra. We shall make use of them very frequently.

11.14.10 Returning to our work on index notation it is the laws of **inverses** which have special importance. The additive inverse of 2 is –2 and x^2 has the additive inverse x^{-2}, which represents $1/x^2$. You can see this in our earlier work on the column values of our ordinary number system.

The multiplicative inverse of x^2 is $x^{1/2}$ which is the **square root** of x. All this leads to some manipulations of indices which need a good deal of practice to develop real confidence. Take:

$$x^2(x^3 + 2x^2 + x + 2)$$

for example, which, when multiplied throughout will become:

$$x^5 + 2x^4 + x^3 + 2x^2$$

11.14.11 We do occasionally meet situations of this kind in modelling real-life problems and a working understanding of index notation is valuable.

11.15 Simplifying

11.15.1 Of much greater importance, however, is the ability to simplify our models and to change their form to suit our needs.

11.15.2 First the simplification process.

At the heart of any equivalence (equation) is the requirement that *we do nothing to disturb the balance* which that equivalence demands. Both sides *must* remain of equal value whatever we do to them. Eventually we would like to see, simply, $x = n$.

11.15.3 The golden rule when simplifying is to remove one complication at a time by performing the *inverse operation* with the *inverse value* on *each side* of the equation.

11.15.4 Let us do so systematically in steps starting with a simple example:

$$\tfrac{1}{2}x + 14 = 19$$

Step 1 $$\tfrac{1}{2}x + 14 - 14 = 19 - 14$$

brings $$\tfrac{1}{2}x = 5$$

Step 2 $$2 * (\tfrac{1}{2}x) = 2 * 5$$

brings $$x = 10.$$

11.15.5 The process works for complicated models also:

	Look at	$7x^2 + 15x - 5 = 5x^2 + 3(5x + 15).$
1.	Remove the brackets:	$7x^2 + 15x - 5 = 5x^2 + 15x + 45$
2.	Remove the '–5'	$7x^2 + 15x - 5 + 5 = 5x^2 + 15x + 45 + 5$
	(additive inverses sum	
	to zero) so it now becomes	$7x^2 + 15x = 5x^2 + 15x + 50$
3.	Remove the 15x	$7x^2 + 15x - 15x = 5x^2 + 15x - 15x + 50$
	or	$7x^2 = 5x^2 + 50$
4.	Remove the $5x^2$	$7x^2 - 5x^2 = 5x^2 - 5x^2 + 50$
5	Divide both sides by 2	$2x^2/2 = 50/2$
	leaving	$x^2 = 25$
6	Find the square roots of	
	both sides	$x = \sqrt{25}$
	Result	$x = \pm 5$

Note: That the square root of any number should, strictly speaking, always be written as, for example, ±5. Only if we know from the circumstances that the value *cannot possibly* be negative are we safe in assuming that any square root is positive.

11.16 Algebra and the machine

11.16.1 The computer, unlike many of *us*, is not intimidated by a complicated model and could deal well enough with the original form of our equation but there are two problems.

There is the inefficiency (which ultimately slows program runs and costs money) of doing tasks which are unnecessarily complicated and, even more important, the greater risk of mistakes or increased error.

11.16.2 A second, but even more important concern, is people-related. *We* and any of our colleagues who may have to work on our programs, *must* be able to understand the models, not just when we have devised them but perhaps years afterwards. It is then that the time taken in simplifying and standardising our models pays dividends.

11.17 Changing forms

11.17.1 It happens very commonly that a model, even one which we have taken from a reference book, may not be in the appropriate form for the application which we have in mind. In that event a certain remodelling is needed.

11.17.2 If, for example, we consult a conventional maths book we shall find that the classic formula for the volume of a sphere is:

$$V = \frac{4}{3}\pi r^3$$

when *V* is the volume and *r* is a *known* radius.

11.17.3 Supposing, however that we need to manufacture a sphere which will contain a specified volume and of which we must determine the radius, which is often the way in which the engineer meets the problem.

11.17.4 We then need to find an unknown *r* for a *known V*, the exact opposite of the maths text book version. In algebraic language we must 'change the subject of the formula' or make *r* the subject of the equation.

11.17.5 We do pretty much the same as the simplification process we have already considered, take one step at a time and use our knowledge of inverses.

1. Remove the 'divide by three' to give $\quad 3V = 4\pi r^3$
2. Divide both sides by 4 $\qquad\qquad\qquad 3V/4 = \pi r^3$
3. Divide both side by π $\qquad\qquad\quad 3V/4\pi = r^3$

4. Find the 'cube root' of both sides $\qquad \sqrt[3]{\dfrac{3V}{4\pi}} = r$

11.17.6 Once again we see a standard algebraic method used to help us prepare our models in the form best suited, for a variety of reasons, to computer use.

11.18 Substitution

11.18.1 The final algebraic method with which we need to concern ourselves is that of **substitution**.

11.18.2 If, at any time whilst we are working on a model, we develop or discover actual *numerical values* for any of the terms we may often simplify matters still further by substituting the value(s) for the unknown(s).

11.18.3 In our most recent example, for instance, the use to be made of the solution will determine the appropriate value for π and when we know the former we should substitute the appropriate number in our model.

11.18.4 If the manufacturing demands are always the same we may be able thereafter to use a **constant** numeric value for π within our program.

11.18.5 Other substitutions may be a little less obvious at first glance. If we have a pair of simultaneous equations:

(a) $7x + 2y = 25$
(b) $x + 3y = 9$

we could, by changing the subject of **(b)** conclude that $x = 9 - 3y$ and substitute $9 - 3y$ for any x in equation **(a)** to give: $7(9 - 3y) + 2y = 25$ and, by our step-by-step process find:

$$63 - 21y + 2y = 25$$
$$63 - 19y = 25$$
$$-19y = -38 \dagger$$
$$y = 38/19$$
$$y = 2$$

($\dagger$ When we have a minus quantity on each side it is often useful to multiply both sides by -1; can you see how this works?)

11.18.6 The traditional manner of solving simultaneous equations might also be said to be a kind of substitution.

In the pair of equations we have just looked at we might substitute a version of equation **(a)** in which the x component matches that of equation **(b)**, namely, $7x + 21y = 63$. We can do this by multiplying all the components of the original equation **(b)** by 7.

If we now set the equations together:

(a) $7x + 2y = 25$ and
(b) $7x + 21y = 63$

we can see that the *difference* between the two equations, found by *subtracting* **(a)** from **(b)** is that

$$19y = 38 \text{ hence } y = 2$$

which is a conclusion we arrived at in the earlier substitution method. In both cases if we substitute this value for y in either equation we shall be able to find the other *unknown*, 'x' and check our results by making the substitution also in the other equation.

11.18.7 What we have done is to substitute for an *inconvenient* form of one of the equations a version in which, whilst better suiting our purpose, *the original equivalence relation is maintained.* We may even make convenient substitutions of *both* equations if it suits our purposes.

11.18.8 Take for example the simultaneous pair:

(a) $3p + 2q = 4$
(b) $4p + 5q = -11$

which will not reveal a usable *difference* by changing one equation only but will do so if we substitute other versions of *both*:

(a) $12p + 8q = 16$
(b) $12p + 15q = -33$

leaving, as the difference

$$7q = -49 \text{ or } q = -7.$$

11.18.9 Check that you understand how the substitute equations were arrived at and try some examples.

11.19 The computer and simultaneous equations

11.19.1 The method is, however, complicated to program for the computer since it involves *looking at* the equations and making *judgements* about them and the relationships between them. Computers are simply no good at inspecting equations and making such judgements.

11.19.2 Fortunately we have yet another long-standing device in the mathematics toolbox in which, rather than substituting another form of the equation(s) we set the values concerned into a matrix and then substitute the **INVERSE** of the **matrix**.

11.19.3 If we look at the first pair of equations in 11.18.5 we can write the whole thing in matrix form as:

$$\begin{bmatrix} 7 & 2 \\ 1 & 3 \end{bmatrix} \begin{pmatrix} x \\ y \end{pmatrix} = \begin{pmatrix} 25 \\ 9 \end{pmatrix}$$

11.19.4 What we have *not* seen hitherto is that not only do numbers and algebraic terms have inverses but so also do matrices, both additive and multiplicative.

For addition

$$\begin{bmatrix} 7 & 2 \\ 1 & 3 \end{bmatrix} + \begin{bmatrix} -7 & -2 \\ -1 & -3 \end{bmatrix} = \begin{bmatrix} 0 & 0 \\ 0 & 0 \end{bmatrix}$$

shows the inverse law.

11.19.5 The multiplicative inverse of a matrix is a little more complicated, but is, happily, freely available in many translators and in virtually all spreadsheets systems. For any 2×2 matrix A we have the model:

$$\frac{1}{A} * A = \begin{bmatrix} 1 & 0 \\ 0 & 1 \end{bmatrix}$$

and we have seen, in an earlier chapter, both the *identity matrix*, with ones in the leading diagonal and zeros in all the other cells, *and* the consequence of multiplying any array by it.

11.19.6 We now need a rule for finding the **inverse** of any 2×2 matrix to solve equations with two unknown. If we label the cells in our 'subject' matrix a, b, c and d, we have:

$$\text{the inverse of } \begin{bmatrix} a & b \\ c & d \end{bmatrix} = \frac{1}{ad - bc} * \begin{bmatrix} d & -b \\ -c & a \end{bmatrix}$$

and it remains only to substitute the *numeric values* for any pair of equations with two unknowns into the inverse matrix in all the appropriate places.

11.19.7 The part *outside the matrix proper*, '$1/(ad - bc)$' is known as the **determinant** of the matrix and when we are finding an inverse by hand we must remember to evaluate the determinant, $(ad - bc)$, and, as a final step, to use its *inverse* to multiply our product matrix.

11.19.8 In spreadsheets and translator programs all these calculations are invisible to us and are 'called' by the **array inverse** command in one form or another.

11.19.9 Let us try our example, multiplying both sides by the inverse matrix

$$\frac{1}{21-2}\begin{bmatrix} 3 & -2 \\ -1 & 7 \end{bmatrix} * \begin{bmatrix} 7 & 2 \\ 1 & 3 \end{bmatrix}\begin{pmatrix} x \\ y \end{pmatrix} = \frac{1}{21-2}\begin{bmatrix} 3 & -2 \\ -1 & 7 \end{bmatrix} * \begin{pmatrix} 25 \\ 9 \end{pmatrix}$$

becomes

$$\frac{1}{19}\begin{bmatrix} 19 & 0 \\ 0 & 19 \end{bmatrix} * \begin{pmatrix} x \\ y \end{pmatrix} = \frac{1}{19} * \begin{pmatrix} 57 \\ 38 \end{pmatrix}$$

then

$$\begin{bmatrix} 1 & 0 \\ 0 & 1 \end{bmatrix} * \begin{pmatrix} x \\ y \end{pmatrix} = \frac{1}{19} * \begin{pmatrix} 57 \\ 38 \end{pmatrix}$$

and finally

$$\begin{pmatrix} x \\ y \end{pmatrix} = \begin{pmatrix} 3 \\ 2 \end{pmatrix} \text{ giving } x = 3; \quad y = 2$$

11.19.10 This is the preferred method of dealing with such simultaneous equations by computer since, once the numbers are properly placed into an array the normal matrix functions do all that is necessary.

11.20 Coefficients

11.20.1 It remains only to add one further comment about *notation*. If you follow up any of these ideas by reading conventional mathematics textbooks the *number values* associated with symbols, for instance the '7' in $7x$ is known, in the jargon, as the **coefficient** of x.

11.20.2 Taking this one step further and looking at $7x^2$, for instance, the actual quantity is made up of *three* parts, the **coefficient**, the **unknown** (value of x itself) and the **index** (or **power**) of the value represented by the x.

11.21 Conclusion

It is in ways like these, which all apply just as much to the **modelling and solution of inequalities**, that classical numerical and algebraic methods can help us with our practical modelling tasks and it is for this reason that at least the techniques and ideas we have looked at in this chapter are worth spending some time on.

Exercises

(1) We can be sure that $1 < \sqrt{2} < 2$ since $1^2 = 1$ and $2 \div 2 = 1$. As a first estimate of $\sqrt{2}$ we could try 1.5.

Squaring this gives 2.25, which is larger than 2 and so 1.5 must be too big an estimate.

We can now say that $1 < \sqrt{2} < 1.5$. As an improved estimate we could now try 1.25 and squaring this gives 1.5625 which is less than 2 and so our new estimate is too small and $1.25 < \sqrt{2} < 1.5$.

Task: Work through this iteration until you can confidently write $\sqrt{2}$ correct to three decimal places.

(2) Write a pseudo-code segment to perform this iteration and test it.

(3) **(a)** Devise an iterative model to solve $y = \sqrt{x}$ where x may take any input value.

(b) How will your model differ if answers are needed to:
   (i) three decimal places or
   (ii) four significant figures?

(4) Find $\sqrt{470}$ by an iterative method.

(5) A vulgar fraction value for π is 22/7 and calculating this gives 3.143 to three decimal places. Use the iteration in 11.7.1 of Chapter 11 to calculate π to three decimal places.

How large is the difference between the two values?

(6) Write the model $P = R * H - T$ with, in turn, R, H and T as the subject.

(7) **(a)** Simplify $7x^3 + 2x^2 - 7 - y = 4x^3 - x^2 - 5$.

(b) If $x \in$ {Integers: x: $1 \le x \le 4$} calculate, for your simplified equation, the corresponding values of y.

(8) Write $y = 2x^2 + 27$ with x as the subject.

(9) Solve the simultaneous equations:

$$3t + 2s = 27$$
$$2t + 5s = 51$$

(a) by 'traditional' algebra
(b) by the inverse matrix method.

(10) Express

$$\frac{(x^5 * x^{-1})}{(x^{-2} * x^4)}$$

in its simplest form.

12

Diminishing uncertainty – statistics

Objectives

After working through this chapter you should be able to:

❏ understand what we mean by the word 'statistics'
❏ see the relevance of statistics to the work of computer staff
❏ perform elementary statistical analyses of sets of data
❏ interpret elementary descriptive statistics.

12.1 Why statistics?

12.1.1 There are many situations in the real world which cannot be modelled by any of the techniques we have considered so far. In the introduction to this book some such problem areas were identified as being those where:

(a) certainty involves destructive testing
(b) the situation is too large or too complex to be measurable with any certainty
(c) we need, on the basis of what has happened in the past, to be able to make useful forecasts of what is *likely* to happen in the future.

12.1.2 Within the organizations which our computer systems serve any one, or more, of these may:

(a) apply to products which the organization manufactures or uses
(b) relate to the organization itself, the country within which it functions, or the markets in which it hopes to compete; and
(c) be involved in the monitoring of performance and profitability which are essential to management *planning* but without which they can only *react*.

12.1.3 All of them will also apply, to a greater or lesser extent, to our computer systems themselves.

12.1.4 The branch of mathematics which covers such modelling problems is generally known as **statistics**.

12.1.5 The word statistics is often used loosely and means many different things to different people but for us it is essential to be very clear about its *precise* nature because all our systems are likely to have to meet demands, from our users, for 'statistical information'. *We* must understand those needs and know how to create or use the software which will produce the information demanded of us.

12.1.6 Within any substantial computer system we shall also need statistical tools to help us evaluate the capacity of those systems, to plan the work-load which they can carry and to appraise the reliability of existing equipment or the many choices available to us when we come to replace or extend any part of our systems. Brochures for computer equipment, even the simplest type of printer, are often full of real or pseudo-'statistical' information and we *must* be able to judge its validity, truthfulness and how useful it is to us.

12.2 The stages

12.2.1 Within what is generally described as 'doing statistics' there are a number of distinct phases:

(1) planning the 'capture' of the raw data
(2) *capturing* the data
(3) organizing that data into usable form
(4) displaying the data in ways which are clear, *true*, and easy to understand
(5) analysing the data by approved methods
(6) deriving *statistics* from the analysis
(7) stating the statistics in approved ways
(8) indicating what conclusions may be drawn
(9) identifying any further assumptions which may be **inferred** from the statistics.

12.2.2 You will notice that in stages from **(1)** to **(5)** I have spoken only of 'data' and that 'statistics' are actually not mentioned until stage **(6)**. This is because *statistics are a finished product* and statistical work is like the information processing cycle where raw data is *input* to a system, *processed* and *output* as information.

12.2.3 In statistical processing the *output* is in the form of statistics whilst the unprocessed raw data are of little use in themselves and can actually be quite misleading.

12.2.4 Strictly speaking, a **statistic** is *the numerical output of a recognized process for evaluating raw data*. People often use the word 'statistics' very loosely in conversation and, worst of all, they say things like 'statistics prove ...'.

12.2.5 This, like so many misinterpretations of the subject, is wholly untrue because statistics alone can never prove anything. They may only *describe* a situation or, at best, provide *evidence* as to whether a given situation seems to be what is to be expected.

12.2.6 What can be *derived* from that evidence, stages (7) and (8), involve more complex tasks which are not our concern at this level. We are concerned with the tools of analysis and reporting, leaving the conclusions and inferences to the user.

12.3 Language

12.3.1 The language of this type of modelling deals with 'tendencies', 'trends' or even 'likelihoods' but *never* certainty; at the very best we can only diminish *un*certainty. This does not imply that we may be casual or careless. We must learn to approach the subject in a strictly disciplined and cautious way but, happily, there is very little really difficult mathematics involved and, as always, when we meet any new symbols or models we will explore and explain them step by step.

12.3.2 We must also remind ourselves that both data and information may be **qualitative** or **quantitative**. The former relating to *attributes*, which we deal with, if at all, either by **sets** or by **non-parametric** statistical methods. In this book we are not concerned with the latter.

12.3.3 *Quantitative* data, which *are* very much our concern, are expressed in *numbers* as we saw earlier, but these, let us remind ourselves, may be on **nominal**, **ordinal** or **ratio** scales and the latter may be **discrete** or **continuous**. For the purposes of this chapter we shall limit our work to the *ratio* scales.

12.4 Sampling

12.4.1 Usually the responsibility for stages (1) and (2), in our list above, will lie with the people who *design* the statistical investigation and our responsibility will be to process the data which have been collected.

12.4.2 What we *shall* need to know is whether the data are from a **sample** or from the whole **population**. For the NCC International Diploma, for example, the *population*, with which any *individual's* performance is compared, is *all* the potential candidates in *all* the schools and colleges in *all* the regions of *all* the countries concerned.

12.4.3 The candidates of any one school will, on the other hand, be a *sample* of that population. For all sorts of reasons that sample may be **typical** or **atypical**, (*not* typical). That is to say that it may or may not be truly **representative** of the population as a whole.

12.4.4 For a truly reliable sample we may have to take our data at **random** from the whole population. There are whole books on sampling methods if you wish to pursue this topic.

12.4.5 It is, however, very important to *us* when we are trying to appraise a manufacturer's claims for the performance of disk drives, for instance, to know how the statistical evidence was gathered. This should include the sample size and the sampling methods if we are to have a reliable picture.

12.4.6 Unrepresentative samples can generate statistics which give a very misleading impression because the data are, by intention or otherwise, **biassed**, 'tilted' towards producing favourable statistics, and therefore must be considered to be defective. Some of the models which we use will need to be modified if the data are from small samples.

12.5 Practical steps

12.5.1 Let us start to model a practical problem as a means of exploring the stages from **(3)** to **(6)**.

A stock-update run is to be fitted into a new work schedule. A sample of 72 runs over several weeks has been timed with the results, in minutes to two decimal places, below:

20.72	14.53	23.31	19.25	16.87	7.15	17.44	12.28	11.30
19.55	23.85	17.14	4.85	10.95	21.65	28.10	11.91	29.64
15.38	11.57	22.61	21.16	13.20	16.03	19.03	22.08	18.39
17.06	24.14	15.19	12.29	19.47	13.39	17.77	23.31	15.16
25.02	16.29	20.03	13.57	16.61	19.41	22.97	18.64	13.12
19.36	12.99	18.73	22.44	14.73	19.83	14.46	19.51	15.82
12.58	26.78	18.28	9.53	20.38	21.30	13.56	17.62	22.80
27.54	10.54	17.59	20.45	23.74	18.70	26.31	11.03	12.41

12.5.2 This is typical of raw data in that it is unordered and almost unreadable as it stands. Before we can interpret such material we need to know if it fits any 'pattern'.

12.5.3 We need to know both the minimum and the maximum times, or the **range**, whether there is some sort of 'average' time and whether most runs are close to that average or are widely scattered.

12.5.4 In formal terms we speak of **measures of central tendency** to describe the 'average' on the one hand and measures of **dispersion** to indicate the degree of scatter. We shall explore these measures more fully as they become relevant.

(Note that *we* only use the word 'average' in a very general way and not to mean anything precise since it has come to mean so many different things in everyday speech that it no longer has any useful meaning at all mathematically.)

12.5.5 With very small numbers of data, fewer than twenty or so items, we may simply **tabulate** and **order** the data.

12.5.6 With larger samples, in order to avoid the sheer volume of work in dealing with all the data separately and because the process often throws up other useful information we group the data into **classes**.

12.6 A distribution

12.6.1 Here, in planning how to organize the data, we need to introduce some new techniques in order to *bring out* the pattern, or **distribution** of the data.

12.6.2 In order to decide on what classes to use we need first to establish the **scale** of measurement and the **range**. In this case the scale is continuous and the range is from the smallest value, 4.85 minutes, to 29.64 minutes. It would obviously be very foolish to schedule 4.85 minutes for every run of the job, but it would also be very wasteful always to allow 29.64 minutes for each run! This clearly *does* not help with the scheduling problem and needs further investigation.

12.6.3 We commonly choose an **odd number** of classes in order to reveal the symmetry of the data if it exists. Ideally not fewer than five and not more than 11, with **integers** as their midpoints in order to minimize error effects. *Our* classes need to cover a range of at least from 4 to 30 inclusive, with a middle value of 17. *Seven* classes each having a 'spread', or **class**

interval, of 4 minutes, would cover a very slightly larger overall range and be convenient to work with.

12.6.4 It is also the case that our data imply a **continuous** scale so that our classes must allow for this by defining boundaries precisely.

As a matter of notation we commonly use x for an individual datum and x_i to identify any *specific* value. We use n for the total number of data.

Our suggested division would give:

$$3 \leq x < 7$$
$$7 \leq x < 11$$
$$11 \leq x < 15$$
$$15 \leq x < 19$$
$$19 \leq x < 23$$
$$23 \leq x < 27$$
$$27 \leq x < 31$$

Note carefully the use of the '$<$' and '$\leq$' symbols and that the use of them unmistakably defines each class as a true set so that there can be no doubt as to the class for which any individual datum is intended.

12.6.5 We now do something which we would think quite improper in ordinary arithmetic models. We assume that all items which fall into a particular class are evenly scattered within that class and that the midvalue can represent any and all of them as a kind of 'average'.

12.6.6 Accordingly we do not record *any individual datum* but merely *count* the number of data which fall into each class. In practice we put these classes into a table, make a mark for each item in a class and gather those marks into groups of five. We have, for centuries, called this 'tallying' items.

12.6.7 Can you write a short pseudo-code segment which could allocate the data to the appropriate classes and perform the 'tallying'?

12.6.8 The number of items in each class, after tallying, is known as the **frequency** of that class and the whole process of organizing the data in this way is described as making a **frequency distribution** of the data. In Figure 12.1 on the next page I have, as the final step, added a column to contain the class frequencies.

12.6.9 To sum up this phase we have made a **frequency distribution** of our raw data by allocating them to seven classes with a **class interval** of four, the midpoint of each class being taken as the value of each item in that class and with the total 'tallies' being the **frequency** for each class.

CLASS	xi	TALLY	fi
3≤ x <7	5	|	1
7≤ x <11	9	||||	4
11≤ x <15	13	⊞ ⊞ ⊞ ||	17
15≤ x <19	17	⊞ ⊞ ⊞ ⊞ |	21
19≤ x <23	21	⊞ ⊞ ⊞ |||	18
23≤ x <27	25	⊞ |||	8
27≤ x ≤31	29	||||	3
		Total = n =	72

Figure 12.1   Tallying the data.

12.7 Displaying the data

12.7.1 We can now put the relevant data into a spreadsheet, for example, and **display** these data in pie charts, bar, or line graphs if all we wish is to see the picture more clearly. The *display* of such material is, for the great majority of people, the best way of making it easily understandable.

12.7.2 Here, in Figure 12.2, is the simple **histogram** of the data and this already tells us a certain amount which is useful.

Figure 12.2   The straightforward histogram.

12.7.3 For example it is clear that *most* of the runs occupy the three centre columns and that a very few took very much less, or very much more time than the majority. It gives a good immediate picture of the data.

12.7.4 We should pause at this point and look at the overall picture presented by the histogram, which may be broadly symmetrical, as in our example, or may be 'skewed' either 'negatively' or 'positively, as shown in Figures 12.3 and 12.4, below.

Figure 12.3   Negative skew.

Figure 12.4   Heavy positive skew.

12.7.5 Symmetry is a norm for many distributions and if we see a histogram with a heavy skew, in either direction, we should call attention to the fact as part of our report.

12.7.6 There may well be good reason for the lack of symmetry which the user will either understand or will wish to investigate further. For our present purposes we should not attempt to *interpret* such results.

12.8 Frequency polygons

12.8.1 For some purposes we may find it useful to develop, from the histogram, a **frequency polygon**. This is constructed by joining the *top centre points* of the columns of the polygon by straight lines. In Figure 12.5 on the next page the polygon has been overlaid on the original histogram so that you can see the relationships.

12.8.2 One attraction of the polygon is that spreadsheet systems can be made to generate such a graph even though few, if any, will actually produce a true histogram.

12.8.3 If, for some special purpose, we use a very large number of classes we may join the points of the polygon by a smooth curve. It is often such 'smoothed' polygons which are shown as models of certain 'standard' distributions.

Figure 12.5   The frequency polygon.

12.9   Cumulative frequency

12.9.1 The information given, either by the histogram or the polygon is, however, primarily **impressionistic** rather than **quantitative** and to this extent the graphs, whilst very useful, are still rather limited.

There is a good deal to be gained by adding another new technique and tabulating the **cumulative frequency**, often written as 'Cu.f', of the distribution, either in **absolute** or in **relative** terms (see Table 12.1).

12.9.2 From our table we can see that one run took *less than 7 minutes*; a total of five runs took less than 11 minutes; 22 runs in all took less than 15 minutes, and so on. This, useful though it may be, would not tell us much if we were

Table 12.1

Minutes	f_i	Absolute Cu.f.
<7	1	1
<11	4	5
<15	17	22
<19	21	43
<23	18	61
<27	8	69
<31	3	72

comparing with a performance plan or with *another* sample of say 97 runs. To overcome this we do what is usual when we wish to compare two or more ratios or proportions and turn to the common denominator of **percentages**.

12.9.3 The *absolute cumulative frequencies* are converted to percentages to produce the *relative cumulative frequencies*. In our present example 1/72 runs took less than 7 minutes; 5/72 took less than 11 minutes; 22/72 took less than 15 minutes, and so on. Converting these to percentages we can both simplify and extend our table as shown in Table 12.2.

Table 12.2

Minutes	f_i	Absolute Cu.f.	Relative Cu.f. (%)
<7	1	1	1.4
<11	4	5	6.9
<15	17	22	30.5
<19	21	43	59.7
<23	18	61	84.7
<27	8	69	95.8
<31	3	72	100

12.9.4 The contents of the table may now be plotted on a graph as shown in Figure 12.6.

Figure 12.6 A cumulative frequency ogive.

12.9.5 This curve does not resemble any of those at which we looked in the chapter on graphs but is known as an **ogive**. The complete graph, plotted from our *relative cumulative frequency* data is known as the **cumulative frequency ogive** of the distribution. In practice we rarely use absolute frequencies for this purpose so you may often find that the 'relative' part of the title has been left out.

12.9.6 If the vertical scale is to 100% its is *relative*, if numeric, as the *actual frequencies* and the maximum is the *n* of the data, it is *absolute*.

12.9.7 As it stands this graph does not tell us much more than the histogram but we *can* read from it our first *measure of central tendency.*

12.9.8 In any distribution the actual **middle value** is often of interest and we call this the **median**.

12.9.9 On the frequency ogive the 50% point identifies the median and if we 'drop' a perpendicular line from where the curve itself is cut by the 50%, horizontal line, we can read off the *value* of the median on the scale of the **variable** ,x, which is 'minutes', in our present case.

12.10 Quartiles

12.10.1 We also often divide a distribution into equal 'slices' of the cumulative frequency. Among the commonest of such subdivisions is to identify the 25%, 50%, 75% and 100% levels of cumulative frequency, dividing our distribution into quarters. We have already seen that the 50% mark leads us to the *median*; similarly the values at the 25% and 75% points on the frequency scale will allow us to read the corresponding points on the horizontal scale.

12.10.2 Values of the *measured variable* which are derived from such regular subdivisions of the frequency scale are called by the rather strange name 'fractiles'. If we divided the frequency scale into sixths, for examples, the corresponding values of the variable parameter would be 'sextiles'. When we subdivide the *f* scale into quarters the resulting values are called **quartiles**.

12.10.3 From this 'annotated' version of the ogive we can now read the *median* and the **first** and **third quartiles**, representing the 25% and 75% frequency 'bands', respectively. This can be seen in Figure 12.7.

12.10.4 We can now say that 25% of runs took not more than 14 minutes and that 25% took more than 21 minutes, with both values being approximate since they are read from the graph.

Figure 12.7 The annotated ogive.

12.10.5 To reach the highest level of approximation in the interpretation of ogives we must, once again, use large scales on accurately printed graph paper and learn to draw the curves as accurately as we possibly can. Nowhere is pride in good craftsmanship better repaid than in the drawing of graphs.

12.11 Dispersion or spread

12.11.1 We can also see that 50% of cases took between 14 and 21 minutes. This is the range between the first and third quartiles and gives us our first useful measure of spread, or **dispersion**, which we call by the rather clumsy name of the **semi-inter-quartile range**, the SIQR, or, more briefly, the **quartile deviation**.

12.11.2 We find this by *halving the absolute difference between the first and third quartiles*. In our current example the SIQR $\approx \frac{1}{2} |14 - 21|$ or 3.5 = the quartile deviation.

12.11.3 In statistics the word 'deviation' has two special meanings. It is either the amount by which any datum differs from the 'average', or it is one of several kinds of *'average' of the spread of individual values about a central point*.

12.11.4 What is important about this technique of classifying the data and drawing the ogive is that, once the range has been allocated to classes there is no further special formula or elaborate model to build. We can simply read off the two measures from the graph.

12.11.5 Moreover the whole technique is ideally suited to working by spreadsheet up to the point of plotting the *points* on the graph. It is better to print out the graph *then* and draw the curve by hand since spreadsheets systems, as we saw earlier, will ordinarily not draw smooth curves.

12.11.6 We can read many other answers from the ogive. Suppose, for instance, the operations manager decided that the schedule, ideally, should allow 20 minutes for this particular job run. How often would it 'over-run'? Look at 20 on the 'minutes' axis, draw a line up to the curve and then across to the frequency scale and we see that approximately 70% of runs would be completed within time and 30% would run over time. He or she might well think that this was an acceptable percentage of overruns. The 'worst case' scenario will be when *all* the jobs happen to over-run on the same shift!

12.11.7 The next benefit from preparing such a graph and deriving the two measures is that by doing so for each of two samples, even of quite different sizes, we can compare medians and SIQRs to see if they are closely similar, or markedly different performances. This is the great importance of **standard measures** of central tendency and dispersion, that we can *compare* sets of data with each other or with **standard models**.

12.11.8 The cumulative frequency ogive is a quite powerful model which is easily derived yet from which we can obtain much useful information. It is, however, a a relatively limited device and there are other and more sophisticated measures which suit more demanding tasks.

12.12 Central tendency

12.12.1 If we look back once again at our histogram we see that there is one column which represents a greater frequency than any other, the $15 \leq x < 19$ class. This class, which contains the largest number of items, is often important and within it lies the **mode**, a second measure of central tendency.

12.12.2 For a reasonable approximation we may take the midvalue of the largest class as the mode but we may also use a slightly more refined technique. In this we take some account of the relative heights of the two adjacent columns by drawing two lines, as shown in Figure 12.8 and dropping a perpendicular to the horizontal scale.

The value indicated by the arrow, at about 17.2 on the scale, is the mode. The graph, as with the ogive, must be drawn as precisely as possible if we are to be able to read such values to a sufficient degree of accuracy to justify the additional complications.

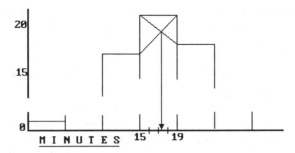

Figure 12.8   The mode refined.

12.12.3 Some distributions may be **bimodal**, which means that they have two modes and appear something like Figure 12.9.

Once again this is an output to which we should call attention but should not attempt to interpret. The user may well expect such an outcome or may, if is unexpected, mount an appropriate investigation. We are not, at the present level, competent to draw the correct inferences.

Figure 12.9   A bimodal distribution.

12.13 Full descriptions

12.13.1 One of the limitations of our *quartile deviation*, as a **statistic**, is that it describes only the middle 50% of a distribution and ignores the outer 50% which is spread over the two 'tails', as they are called. Since there may be considerable variation between the tails of this and any other distribution there will be many purposes for which we need more detailed information which takes account of the *whole* of the data.

12.13.2 We need to look at a more detailed description (and what we are dealing with are all known as **descriptive statistics**, often accessed in statistics software packages by the command **describe**). It is most convenient to do this by taking a simpler data set, to keep the arithmetic to a minimum, because we shall be using some new techniques and building a new model which can then be developed for the more elaborate data sets.

12.13.3 We need, let us remember, a measure of *central tendency* and a measure of *dispersion* for a data set like: 12, 9, 7, 18, 5, 14, 11, 19, 15, 8. This is clearly too small to put into classes and so we deal with it in a different way. First we find the *third* of our common *measures of central tendency*, the **arithmetic mean**, by adding all the *x*s and dividing by the number of items. The three are the **median, mode** and **arithmetic mean**. Each is arrived at by a different procedure, describes different aspects of the distribution and is used for different purposes. This is why we treat the word 'average' with such caution. There are at least *five* commonly used measures of central tendency and what the layman means by 'average' might be any one of them!

12.13.4 This is an opportune moment to meet a very common but rather forbidding-looking use of symbols. We may well find, in a mathematics or statistics textbook, the *arithmetic mean* indicated by:

$$\bar{x} = \sum_{i=1}^{i=n} x_i \div n$$

Let us take this, symbol by symbol and translate it into plain words.

The symbol $\bar{x}$ designates the **arithmetic mean** of all the values of *x* whilst the *n* is the number of data items, which we have met before. The unfamiliar Greek upper-case 'S', the Σ symbol, means 'add together all the values' of whatever is to its right, in this case x_i, the individual values of our data variable.

Underneath the Σ is $i = 1$ which means *from* x_1 whilst *above* the Σ is $i = n$ which means *to* the *n*th value of *x*, or in other words the symbols are mathematical shorthand for, 'the arithmetic mean (of a set of data) is found by adding together all the values of *x* from the first to the last and dividing that total by the number of data items'.

This is all that the elaborate picture really means, but it is valuable because it saves quite a lot of words, is completely clear to any mathematician or statistician anywhere in the world, and is wholly unambiguous. Like so much mathematical symbolism it is a very powerful shorthand and the use of such symbols is especially common in statistics.

In practice the 'from' and 'to' parts of the notation are often left out and it is written as 'Σx_i' but the meaning is exactly the same. If, however, we wish to sum, (add together), only *part* of a data set we *must* use the full version.

12.13.5 For brevity, once again, we commonly use only the word 'mean' when we are speaking of the *arithmetic mean* but we must be extremely careful here since there are several 'means' in use. Perhaps we should agree that we will use 'mean' *only* when we intend the *arithmetic* version and that we will use the full name of any other 'means' which we may encounter.

12.13.6 Once we have established the mean we can calculate the **deviation** of each datum, that is to say *the amount by which it differs from the mean*, which we usually label *d*. We must recognize that some values will deviate to the *lower* side of the mean and some to its *higher* side. Those below will, then, have negative values whilst those above will be positive. It is a good idea always to show the sign of the individual deviation since it is often a matter of considerable importance. Being below 'average' is *very* different from being above.

12.13.7 Let us prepare our data table so far, as show in Table 12.3.

Table 12.3

$n = 10$	x_i	d_i
	12	+0.2
	9	−2.8
	7	−4.8
	18	+6.2
	5	−6.8
	14	+2.2
	11	−0.8
	19	+7.2
	15	+3.2
	8	−3.8
Σ	118	0.0
Mean $\bar{x}$	12.8	

12.13.8 To find the 'average' of the deviations is clearly impossible since $\Sigma d_i = 0$. This is inherent in the very nature of the arithmetic mean and will always be the case.

In order to make all the deviations into positive quantities, so that we *may* find a mean, we resort to a very old mathematical trick and *square* each of the deviations. If we then find Σd_i^2 we *can* find *its* mean and finally take the *square root* of that mean in order to arrive at a true mean of *all* the deviations, not just half of them as in the *mean deviation*.

12.14 Standard deviation

12.14.1 Some of you, if you have done much work in electrical or electronic theory, may recognize this process as finding the *root mean square*, or RMS of a set of data. For our purpose in statistics we call it the **standard deviation** and to describe it we use either the abbreviation SD or the Greek letter σ. Usually we shall see the formula:

$$\sigma = \sqrt{\frac{\sum(x_i - \bar{x})^2}{n}}$$

Our amended data table is shown in Table 12.4:

Table 12.4

$n = 10$	x_i	d_i	d_i^2
	12	+0.2	0.04
	9	−2.8	7.84
	7	−4.8	23.04
	18	+6.2	38.44
	5	−6.8	46.24
	14	+2.2	4.84
	11	−0.8	0.64
	19	+7.2	51.84
	15	+3.2	10.24
	8	−3.8	14.44
Σ	118	0.0	197.6
Mean $\bar{x}$	12.8		$\sigma \approx 4.45$

12.14.2 In this case we find that the *mean square* of the deviations is 19.76 and this value is often used in statistics and is called the **variance** or σ^2.

12.14.3 Here we meet a new problem and a new remedy. It has long been established that small samples, below about 40 items (some authors suggest 50) tend to produce results with substantial inherent error. In order to diminish this we apply a correction by using, for finding σ or σ^2, $n-1$ rather than n in all our calculations so that the formula becomes:

$$\sigma = \sqrt{\frac{\sum (x_i - \bar{x})^2}{n-1}}$$

which would give, for our present set of data, a *variance* of 21.96 and a *standard deviation* of 4.69.

12.14.4 It is not usual to do statistics with very small *populations* and it is wise to use the $n-1$ version of all statistical formulae when dealing with situations where $n < 40$. In many texts such small data sets, such as we have just looked at, will be called 'ungrouped' data whilst data which have been set into classes are often referred to as 'grouped' data.

12.14.5 Let us now look again at our set of grouped data and see how that affects the calculations of $\bar{x}$ and σ.

For our variation on the SD model we change the detail of the tabulation to avoid having first to find $\bar{x}$ and the individual deviations. As you will recall, we take the mid-value of the class as the x for all members of that class. Our new tabulation provides the means of using that simplified procedure (see Table 12.5).

Table 12.5

x_i	f_i	x_i^2	$f_i * x_i$	$f_i * x_i^2$
5	1	25	5	25
9	4	81	36	324
13	17	169	221	2873
17	21	289	357	6069
21	18	441	378	7938
25	8	625	200	5000
29	3	841	87	2523
	$n = 72$	Σ	1284	24752

12.14.6 First let us see a different arrangement of the model which we used earlier to find σ. It looks more complicated but is not really greatly so

$$\sigma = \sqrt{\frac{\sum (f_i x_i^2)}{n} - \left(\frac{\sum (f_i x_i)}{n}\right)^2}$$

Within this model we find, inside the right-hand pair of brackets, the mean, which we earlier did manually. This value, when squared, is subtracted from the mean of the *total* x^2 values, on the left.

Substituting in this formula the values in our table, above, we find that this gives $\bar{x} \approx 17.83$ and $\sigma \approx 5.07$ for the full data set.

12.14.7 Compare these with the **median** and **SIQR** which we found by the cumulative frequency model earlier. Can you account for any differences there may be?

For ease of manipulation this is by far the simplest model since, from the time of deciding the classes and their midpoints, and the subsequent tallying of the data to produce the frequencies, all is simple arithmetic, (if rather tedious and subject to human mistakes).

The model may, with a little care, be used in a spreadsheet system without amendment and the spreadsheet system can perform all the tedious arithmetic. If we need to use it at all frequently it lends itself ideally to being manipulated by a *macro* written in the meta-language of the spreadsheet. It is not even too difficult to write a program segment to perform the task although there are many suitable packages on the market.

12.14.8 The mean, median, mode and standard deviation are the major **descriptive** or **summary statistics** offered by any good statistics software package. The choice as to whether to use the tabulation/cumulative frequency ogive technique, or the *calculated values* which we have just been looking at will depend upon what use is to be made of the statistics and whether speed is more important than detail.

12.15 Error

12.15.1 Before leaving these descriptive statistics there is a last important concern. Virtually all numbers have inherent error, as we saw in Chapter 5. All our raw data, then, will have such error and the *statistics derived from them will also have error.* Not only are $\bar{x}$ and σ subject to induced error, rounding error and the like, but they will be affected by the accumulated error within the data and induced by the processing.

12.15.2 We need not go into the calculations which lead to the concept of **standard error** but we do need to know how to calculate and publish such results from our processing. The two main ones, for our purposes are:

(a)   the **standard error of the mean**

$$SE_{\bar{x}} = \frac{\sigma}{\sqrt{n}}$$

(b) the **standard error of the SD**

$$SE_\sigma = \frac{\sigma}{\sqrt{2n}}$$

12.15.3 We should append these to the reported $\bar{x}$ and σ, respectively when we make our final printout unless otherwise instructed.

12.16 Trends

12.16.1 One very valuable tool to our users is a is a **trend analysis** and we may well be asked to produce the analysis of data which will allow the necessary graph to be generated. Typically the data will be describing in a way similar to Table 12.6.

Table 12.6

Year	First quarter	Second quarter	Third quarter	Fourth quarter
1988	32.0	32.7	32.5	33.9
1989	35.3	34.2	34.0	35.8
1990	37.5	35.9	36.8	38.2
1991	39.4	37.9	38.3	37.9

There are, it is clear, many fluctuations and we may see these on any ordinary graph. What we need, however, is an indicator of whether the variations are **random**, and therefore **unpredictable** or whether there is some underlying pattern.

12.16.2 One tool which we use in modelling this situation is known as the **moving average**, (in this case the 'average' is the *arithmetic mean*).

12.16.3 If the data are in true sequence of **time** as in this case, the collection (32.0, 32.7, 32.5, 33.9, 35.3, 34.2, ... , 37.9) is known as a **time series**. (In strict mathematical terms it should be known as a 'sequence' *not* a 'series' since it does not obey the rules governing the latter. However, the name is the common one and we must accept it.)

What follows, however, is that both *time* and *stock level*, in this case, are variables, each seemingly on a discrete ratio scale but with an *implied continuous scale*, see Table 12.7.)

This is to say that, although the stock was actually valued at specific points on the time scale and appears to have discrete values, the *actual* stock value

Table 12.7

Year	Quarter	s_i	$\bar{x}_i$
1988	1	32.0	
	2	32.7	32.35
	3	32.5	32.42
	4	33.9	33.16
1989	1	35.3	34.23
	2	34.2	34.22
	3	34.0	34.11
	4	35.8	34.95
1990	1	37.5	36.23
	2	35.9	36.06
	3	36.8	36.43
	4	38.2	37.32
1991	1	39.4	38.36
	2	37.9	38.13
	3	38.3	38.21
	4	37.9	38.06

will have been varying on a day-by-day basis and between any two quarters will probably have fluctuated considerably.

12.16.4 Since we now have *two* values which are changing, as compared with one variable and *fixed* class intervals in earlier models, we have a **bivariate** distribution as distinct from the **univariate** data which we have seen hitherto in this chapter.

12.16.5 Table 12.7 shows what develops if we 'map' the series contained in our data table in 12.16.1, above, onto the 'moving average' by the model: if s_i = any individual stock level and $\bar{x}_i$ is any individual average then:

$$\bar{x}_1 = \tfrac{1}{2}(s_1 + s_2); \ \bar{x}_2 = \tfrac{1}{2}(\bar{x}_1 + s_3); \ \bar{x}_3 = \tfrac{1}{2}(\bar{x}_2 + s_4)$$

and so on. Note that, in order to minimize error propagation, the $\bar{x}$ values are to two decimal places and that there is no $\bar{x}$ for the first quarter of the time series.

12.16.6 We may now plot the data on a graph (shown in Figure 12.10) and since the 'average' may be assumed to be moving *continuously*, not in *discrete steps*, as are the stock figures, the latter should be shown as a *broken*, the former as a *continuous*, smooth line. If we wish we may now attempt to draw a **line of best fit**, that is to say a line which, by eye, seems to 'average' the continuously varying *moving average* line on the graph. I have drawn in such a line in the graph of our data and we see three distinct patterns. First we see the quite large fluctuations in stock value on the quarter-to-quarter basis, shown by the broken lines.

Figure 12.10 A time-series graph extended.

12.16.7 We next see the 'smooth' variation of the moving average as a continuous but still fluctuating function, but varying much less than the 'point' stock values. Finally we have the straight line on the graph, which is an attempt to interpret the moving average fluctuations as an overall **trend**, by drawing a line of best fit from the moving average points.

12.16.8 Although estimated this gives a useful indication. We must not, however, set too much importance on what we might *extrapolate* from this trend for the future since we have no evidence upon which to make any confident predictions.

12.17 Scatter

12.17.1 The final type of analysis of data which concerns us here is where we have two variables but where it is reasonable to assume that one *might be* a cause of some variation in the other. This is often described as a *causal relation-ship* or a *causal connection* between the one variable and the other.

12.17.2 In such circumstance we call the one which we believe might be the cause the **independent variable** which we always plot on the *horizontal axis* of a graph with the other, the **dependant** variable, plotted on the vertical axis.

We may assume, for example, that people's weights depend to some extent upon their heights. It is clear that if this is so it is not the greater *weight* which causes people to grow taller. It might, on the other hand, be reasonable to suppose that as we grow taller our weights also increase. In this case the *independent variable would be height* with weight being the dependent. We do need to be a little cautious when deciding between any two variables and we should always pause for thought.

12.17.3 Suppose we assume that the sale price of shops which deal in the same goods will largely depend upon their weekly takings and that a sample gives the following data shown in Table 12.8.

Table 12.8

Shop	Takings $00 ($x_i$)	Sale price $000 ($y_i$)
A	10.0	25.6
B	6.0	12.4
C	5.5	10.2
D	8.0	14.0
E	16.5	38.5
F	7.5	13.0
G	12.7	27.0
H	15.0	34.8
I	9.5	14.0

We can tell little from looking at the table other than that there seems to be a *tendency* for the price to rise as the takings get larger. To be rather more sure we need to see the picture more clearly. Whenever you find yourself thinking like this it is well worth while to try plotting a graph.

12.17.4 In this particular case the individual values of the variables, whilst on continuous ratio scales, are actually not connected to each other since each shop is quite a separate entity from all the others. We can only, in this situation, plot a **point graph**. For each shop we plot the point at which the takings and the price correspond and we may, as in our example, label each of the points.

12.17.5 What we now look for is any *pattern* to the 'scatter' of the individual points.

12.17.6 In Figure 12.11 the points all lie close to the 'line of best fit' which has been put in 'by eye'. This suggests that the relationship between takings and the sale price of these shops *are* very closely linked, that they 'relate' to each other and we call such a correspondence the **correlation** between the data.

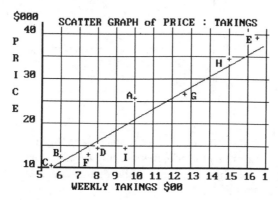

Figure 12.11   The scatter graph.

12.17.7 Such scattered points may be quite clearly random, as in Figure 12.12, or may congregate within a circle and in such cases we may see that there is no correlation. The variables are not linked by some 'cause and effect' relationship.

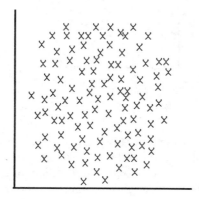

Figure 12.12   No pattern.

12.17.8 If, on the other hand, the points fall within a close ellipse which slopes *upwards from left to right* it indicates a strong *positive* correlation indicating that as the *x* value grows, so does the *y* parameter and that the correspondence between them is very close. The 'fatter' the ellipse the weaker is that relationship.

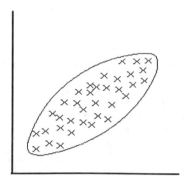

Figure 12.13   A strong positive correlation.

12.17.9 It is not likely to come as a surprise that a 'tight' ellipse which slopes from top left to bottom right indicates a strong *negative* correlation. This means simply that the *y* parameter *decreases* as the *x* parameter *increases*. This is shown in Figure 12.14.

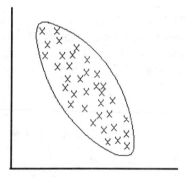

Figure 12.14   A strong negative correlation.

12.18 Conclusion

Once again we find powerful, ready-made techniques and procedures from the mathematical toolbox but we should be a little careful when we use them. We are, at this level, concerned with the *modelling* of the different statistical situations and should be very wary of inferring too much from the results. That is a later stage of statistical analysis.

Exercises

(1) What is the arithmetic mean of {Integers, x: $0 \leq x_i \leq 50$}?

(2) Here is a data table:

Age	Frequency
26–30	5
31–35	43
36–40	66
41–45	91
46–50	87
51–55	105
56–60	83
61–65	80
66–70	25
71–80	14
>81	1

(a) Re-allocate the data to five groups, £30; 30 – 40; 40 – 50; and so on. Show the data as:
   (i)   a pie chart;
   (ii)   a proportional bar chart; and
   (iii)   a histogram.
(b) Using the original tabulation:
   (i)   create a cumulative frequency table;
   (ii)   draw a relative cumulative frequency ogive; and
   (iii)   state the median, mode and SIQR.
(c) Calculate the mean, variance and standard deviation.

(3) The turnover of a small shop in five successive months is: $18400, $17800, $21200, $17000, $15800. Can we assume, on the basis of the last three month's figures, that the business is declining?

(4) An examination showed the following distribution of marks:

Marks	Frequency
0 < 10	3
10 < 19	5
19 < 28	5
28 < 37	8
37 < 46	12
46 < 55	17
55 < 64	24
64 < 73	17
73 < 82	6
82 < 91	2
91 – 100	1

(a) Draw the histogram.

(b) Find the mean, mode and standard deviation.

(c) There is a rough rule that in such circumstances 2/3 of candidates should have marks which fall within ±1 SD of the mean. Does that hold good here?

(d) A similar rule suggests that about 4% of candidates should have marks which differ from the mean by more than twice the SD. How do these data compare?

(5) A commodity traded in both Singapore and London showed the following average prices over an eight-month period:

Singapore $	83	85	88	80	95	98	97	94
London £	28	29	29	27	31	32	32	32

(a) Plot the data on a suitable graph.

(b) Is there any sign that the prices in the two places are related to each other?

(6) 59½, 60½, 61, 61½, 62, 62, 62½, 62½, 63,
63, 63, 63½, 63½, 63½, 64, 64, 64, 64,
64½, 64½, 64½, 64½, 64½, 65, 65, 65, 65,
65, 65½, 65½, 65½, 65½, 65½, 65½, 66, 66,
66, 66, 66, 66, 66½, 66½, 66½, 66½, 66½,
66½, 66½, 67, 67, 67, 67, 67, 67, 67,
67½, 67½, 67½, 67½, 67½, 67½, 68, 68, 68,
68, 68, 68, 68½, 68½, 68½, 68½, 68½, 68½,
69, 69, 69, 69, 69, 69½, 69½, 69½, 69½,
69½, 70, 70, 70, 70, 70½, 70½, 70½, 71,
71, 71, 71½, 71½, 71½, 72, 72, 72½, 73,
73½, 74½

(a) Organize these data into an appropriate distribution and derive all the common descriptive statistics.

(b) Produce any graphs which you think would help present a clear picture of the data.

13

Will it work? Probabilities

Objectives

After working through this chapter you should be able to:

❏ understand the fundamental ideas of probability
❏ use probability models to evaluate computer systems
❏ model probabilities in a variety of ways
❏ select models to suit specific needs.

13.1 Probability modelling

13.1.1 The world is full of *uncertainties*: there are in fact very few total certainties. Despite this we have to plan; we have to evaluate, and allow for, risks; we have to compare the reliability of systems and we have to attempt to forecast what is *likely* to happen.

Mathematicians have, over several centuries, developed and studied **probability models**. It is to these that we turn in order to represent an uncertain world and to satisfy the practical demands of living in it. In this respect the subject matter of this topic is closely allied to that of the previous chapter.

13.1.2 We, as practising computer staff, need to have a grasp of both the principles and the practice of the application of probability models and theories for two reasons. Firstly our users, especially at management level, will need to make use of probability models for much of their planning and forecasting.

13.1.3 Secondly, at both the practical and the theoretical levels, our own computer systems are 'complex states', as referred to in the introduction to this book. They can only be evaluated by statistical methods and their behaviour forecast by probability models.

13.2 Language

13.2.1 There is, as always, some new language to learn but little in the way of new symbols.

The first word to have a special use is **event**, to refer to anything which has happened in the past or may happen in the future. It is the all-purpose word for everything from the toss of a coin to the printer going down in the middle of a run. The former is so trivial that we might hardly consider it worth calling an 'event' in real life; the latter may have such an effect on the workplan as to be seen as a catastrophe! Nevertheless, for our present purpose, each is simply an 'event'.

13.2.2 The second commonly used word is **outcome**, to mean the result of an experiment or of a *sequence* of events. Translated into everyday speech it means something like 'final result'.

13.2.3 A third word is **trial**, which indicates any individual test or example within a series of tests.

13.3 The probability 'scale'

13.3.1 A principle of major concern, from the outset of this study, is that the **probability** of any **event** occurring in the future is **evaluated on a scale from zero to one**.

13.3.2 The **totally impossible** event (although it is very difficult to make a list of examples since scientific progress has a habit of making yesterday's certainties obsolete), has the **probability 0**.

13.3.3 Any **absolutely certain** event, again very difficult to list, has the **probability 1**. Most events will have a probability, p, of $0 < p < 1$.

13.3.4 By whatever numerical process we evaluate past events and from them calculate what is likely to happen in the future our *'probability' answers cannot possibly be less than zero or more than one.* Many, many marks are lost in examinations when candidates forget this basic rule.

13.4 A simple model

13.4.1 At the simplest we may look at a *single* event and evaluate it. Suppose that our printer, according to the records of the past year, has a mean down-time of 20 minutes per eight-hour shift, what is the probability that we shall get through a full shift with a working printer?

13.4.2 First we must compare like with like, which means that the same *set* or the same *measuring scale must be used.* Minutes are *not,* in the strict sense, *on the same scale as, or like,* hours so we must state the recorded mean as 20 minutes out of any 480 minutes.

The probability that the printer *will* break down is 20/480 and the probability that it won't is 460/480. We often use *p* to denote the probability that the event in which we are interested *will* take place and *q* to indicate the probability that it *will not.*

13.4.3 The first 'law' of probability is that $p + q = 1$ and *nothing in any individual situation can change this.* If we should ever arrive at any other result our conclusions are simply wrong. In the present example, if we first reduce our fractions to 'lowest terms' we see that a breakdown has:

$$p = 1/24 = 0.0416 = 4.2\%$$

whilst a shift unhindered by printer breakdown has:

$$q = 23/24 = 0.9583 = 95.8\%$$

and $p + q = 24/24$, 1.0 and 100% respectively.

The different ways of writing the fraction are all valid and equally commonly used so we should use the form which best suits the data.

13.4.4 Note that here that we must not confuse probability with 'odds'. The odds *against* the printer breaking down, i.e. that it *will not,* are 23 to 1.
Be sure that you can see why.

13.4.5 These statements of probability are mathematically accurate but they do not, of course, in any way *influence* the *actual* behaviour of the printer, although this appears to be the belief of many devoted gamblers when it comes to betting! Nor do they indicate what *will* happen on any *particular* shift; they refer only to the *risk* of breakdown.

13.4.6 We deal, not in predicting certainty but in forecasting what is *likely* to happen. The operations manager or shift leader must either *be able to* *evaluate* the true risks, and take account of them in planning the workload, or just trust to luck that the system will be adequate.

13.5 Probability spaces

13.5.1 In a more complex situation it is often useful to set out a **Venn** diagram to model a **probability space**. We can then identify *all* the elements and include them in the universe of discourse. Let us consider an example.

In a computer department of 120 staff there were 15 systems analysts, 40 programmers, and 56 operators. Of these, however, 24 of the operators were also competent programmers and two were also trained analysts; ten of the programmers were also trained analysts and one person was competent in all three skills.

13.5.2 From this information we can construct a model which embodies all possible dispositions of skills (see Figure 13.1). Note that the regions in the diagram have been simplified as in Section 3.6 of Chapter 3.

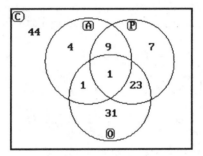

Figure 13.1   One form of probability space.

13.5.3 The Venn diagram allows us to list, by direct observation, *all* relevant initial probabilities.

In the examples below we can list the probabilities as to what skills any member of staff, chosen at random, possesses.

13.5.4 Let us call our unknown person X. The probability that he or she will be a programmer, an element of set *P* is:

$$40/120 = 1/3 = 0.3 \text{ recurring} = 33\tfrac{1}{3}\%$$

By using formal set symbols we can complete a list:

$$p(X \in A) = 15/120 = 1/8 = 0.125 = 12\tfrac{1}{2}\%$$

which reads, 'the probability that X is an analyst (a member of the set of analysts) is one out of eight'.

$$p(X \in O) = 56/120 = 7/15 = 0.46 = 46\frac{2}{3}\%$$

$$p(X \in (A \cap O)) = 2/120 = 1/60 = 0.016 = 16\frac{2}{3}\%$$

$$p(X \in (P \cup A)) = 45/120 = 3/8 = 0.375 = 37\frac{1}{2}\%$$

$$p(X \notin (A \cap O)) = 118/120 = 59/60 = 0.983 \text{ recurring} = 98\frac{1}{3}\% \text{ and so on.}$$

Make sure that you can put each of these into words.

13.5.5 Can you complete the table for all *eight* domains of the diagram?

13.5.6 It is clear that we could also identify the probabilities for all the various combinations of *complements* and subsets and you should try a number of these.

13.5.7 There is *no* probability concerning the known data which are not retrievable from the diagram. If we, for instance, reverse the question and say 'what is the probability that X will *not* be a programmer?' we have a ready-made procedure. Since *all* the probabilities must total one the answer to our new question is $1 - p(X \in P)$ and this is $1 - 40/120 = 80/120 = 2/3 = 0.6$ recurring $= 66\frac{2}{3}\%$.

13.5.8 If we take the histogram of the distribution of job-runs from the previous chapter, even this allows us to assess probabilities.

If, for example, we need to know how likely it is that any job taken at random will be completed in 13-minutes or less we can see that of the 17 jobs in that class half, or $8\frac{1}{2}$, are likely to be below the 13-minute point to add to the 5 in the two lower classes to give a total of $13\frac{1}{2}$ out of a total of 72 job runs in all which will take $\approx$13 minutes.

This is easiest to deal with as:

$$27/144 \text{ or } 0.1875 \text{ or } 18\frac{3}{4}\%$$

which is the probability which we are seeking (see Figure 13.2 overleaf).

13.5.9 You may already have concluded that the smooth curve of the cumulative frequency ogive, if well plotted and drawn, will actually give us many more answers than the histogram. Check this statement by carefully plotting the curve for yourself then taking various *times* and evaluating the associated probabilities. Now look at Figure 13.3 on the next page.

Figure 13.2   The histogram as a probability space.

Figure 13.3   The ogive.

13.5.10 In the last chapter the 'tails' of a distribution were mentioned and by looking at both tails of the graph we can answer such questions as: 'what is the probability that any given run will be *more or less* than five minutes "outside" the median?'.

If we look at our ogive we can see that we are referring to all the runs *below* 12 minutes and *above* 22 minutes. If we now read the relevant percentages included in these tails and *add them together* we shall have the *total* probability which we seek. If your answer is *approximately* 30% you have grasped the principle; any other answer and you should look again. You should also be able to give good reasons why the answer can only be an approximation. Could you give any indication of the inherent error?

13.6 More detailed spaces

13.6.1 The question in 13.5.8, about this distribution, involved a **one-tail**, that in 13.5.10 is a **two-tail** test.

13.6.2 The usefulness of such tests of probability within the probability space represented by a frequency distribution highlights, once again, the difficulties of comparing *absolute* statistics.

13.6.3 If, for example, we wish to compare the performances of a person whose marks in computer studies are 67% and in programming are 52% the first result *appears* clearly superior. This is quite deceptive since what we do *not* know is the relevant central tendencies and spreads of *all* the candidates in *both* subjects. If it turned out that the *statistics* were:

Computer studies	$\bar{x} = 55; \quad \sigma = 14$
Programming	$\bar{x} = 42; \quad \sigma = 8$

we can see that the *overall* patterns of the two sets of results were very different and it is difficult to see whether either of our candidate's performances is better or worse than the other.

13.6.4 *Theoretically*, if a $1 - 100$ scale is being fully used, we should expect the mean to be 50% and the SD 14 marks.

However, different numbers of people sit different tests with different examination papers. Different markers are subject to all sorts of human differences and so are the people who sat both tests in our example. So many differences in circumstances lead to differences in the statistics, such as those in our example, being commonplace. Those same differences *do*, however, prevent any meaningful comparisons on an *absolute* basis.

13.7 Standard scores

13.7.1 We must therefore find ways of making *relative* comparisons and in order to be able to do so on a useful basis we have the concept of the **standard**

score (or **measurement**), denoted by the symbol z. We find z for any individual data item, x_i, by the formula:

$$z_i = (x_i - \bar{x})/\sigma$$

13.7.2 Calculate the z scores for our candidate's two performances. Your results should be:

| Computer studies | z = +0.86 |
| Programming | z = +1.25 |

In both cases his or her performance was above the mean but the performance in Programming was actually better, *compared with the overall performance of all candidates*, than was the performance in Computer studies.

13.7.3 When we are seeking probabilities of any given event within a *large* population we commonly use a probability distribution model which is a smoothed frequency polygon scaled in **standard (z) scores**.

13.7.4 Figure 13.4 is shows the curve of the 'normal' distribution, which is characteristic of a great many situations, such as human heights – or performance in computer programming examinations, if the total population is *very* large.

Figure 13.4 The 'normal' curve.

13.7.5 The graph has been subdivided into one- and two-tail versions of the same z scores scale and the statistics around the perimeter of the curve show the proportion of cases outside and including that particular z score. We can see that, for example, approximately 16% of cases will have z scores of more than one (standard deviation *above* the mean). If we look for those who have scored more than one SD *less than the mean* we find that this is also approximately 16% of the total. Each of these, *individually,* is a one-tail test.

13.7.6 If we ask for *all the cases* which differ from the mean by at least one SD, whose z scores are ≥1, as a complete group both above *and* below the mean, we shall find the answer on the *two-tail* side of the graph and it is, not surprisingly, 32%, double that for the one-tail part.

13.7.7 From this pattern we can establish the *probability* of any individual z score arising. If, for example, we look back at our candidate's programming test performance, which had a z score of +1.25, we can see, on the one-tailed side of the curve, that such a z score describes somewhere between 16% and 6.7% of candidates.

13.7.8 To find out *just what that percentage is* we may use the technique of **linear interpolation**. 1.25 is half-way between 1 and 1.5 so the percentage of cases concerned will be *half-way* between 6.7 and 16, namely 11.35. This means that only 11.35% of candidates, in a normal distribution, would reach or exceed a z score of +1.25.

The performance in computer principles, on the other hand with a z score of +0.86 represents some proportion of the population between 0.5 and 1, or between 31% and 16% of the population. The difference in z scores is 36/50 of the gap between 0.5 and 1 so the value we are seeking will be 36/50 of the gap between 16% and 31%: $31 - 36/50(15) \approx 20.2\%$. The performance reached by *that* z score is that which might be expected of that proportion of the population.

13.7.9 Rather than going to the trouble of *calculating* such interpolations, if we are likely to be much concerned with statistical probabilities, we shall find a table of z values and the corresponding proportions of the normal distribution in any statistics textbook or good mathematical tables.

13.7.10 Putting these results into the form of *probabilities* is now very straightforward since the probability in each case is the same as the proportion of the population which might be expected to reach that score.

13.7.11 On the graph have been marked the 5% and the 1% **probabilities** of any 'one-tailed' event.

13.7.12 By the process of reading the graph 'in reverse' so to speak we find that there is only a 5% probability of any such event having a z value of ≥1.65 and only a 1% probability of it having a z value of ≥2.33. We can also use this information the opposite way round, so to speak.

13.7.13 On the two-tailed side there is only a 5% probability of events falling into the combined tails above or below z = 1.96. This means that for a large number of events which might be expected to conform to this 'normal' distribution the probability is that 95% will fall within the band of ±≤1.96 SDs from the mean.

13.7.14 The normal curve is, like any other frequency distribution, another kind of probability space within which are to be found all the evaluations which we might wish to use.

13.8 Multiple probabilities

13.8.1 So far we have looked only at *single* events but many of the situations with which we have to deal involve a *number of elements* which may influence outcomes.

A typical computer system, for example, will incorporate a processor and a number of peripherals. Each of these elements will have its own speed of data transfer; its capacity for handling or storing data; as well as its rate of failure, both as we have logged them in action *and* as the manufacturer's *claimed* parameters for all of these attributes.

In planning the procurement of systems, in evaluating their *actual performance* and what workloads may consequently be planned for, we are, inevitably, dealing with probabilities and statistics.

13.8.2 Before considering actual cases we had better spend a little time looking at simplified situations so that we may establish some rules.

13.8.3 Every individual event is, in one sense, completely new. We sometimes sum this up in the saying 'a coin has no memory.' This means that no matter how many times we toss a coin every *new* toss is a completely fresh start.

13.8.4 In any sequence of trials, however, events may be **independent** or **combined**. In the former case any one event has no influence on or relation to any subsequent event. In other cases we may be concerned with quite complex *trains* of events.

13.8.5 Every single independent event has a determinable probability for both *p* and *q*. In the sake of a coin the probability of it coming down 'heads' is $p = \frac{1}{2}$ and *not* coming down 'heads' is $q = \frac{1}{2}$.

13.8.6 When we consider events which may be *combined* we take their independent probabilities and combine them arithmetically in one of two ways. If the events are combined in an AND relationship the independent probabilities are **multiplied** to find the outcome. If the association of events is by an OR relationship the independent probabilities are **added**. Initially these may be represented by BOTH and EITHER respectively.

13.8.7 We may see, from an earlier example, the BOTH and the EITHER *relationships* although, since the members are already allocated to their sets, no further arithmetic is needed.

In Figure 13.5 we see, shaded, the ten people who figure in the *A* and *P* set.

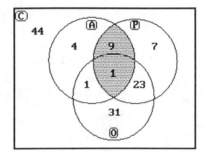

Figure 13.5   In *both A* and *P*.

In Figure 13.6 it is the people who are *either* in *A* or *P*. We should note particularly that because of the intersections the *number* of people concerned is not 15/120 + 40/120, but 45/120.

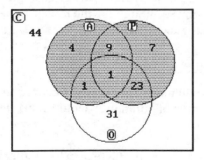

Figure 13.6   In *either A or P*.

13.8.8 To look first of all at just the arithmetic involved let us consider the case of a computer system which used two disk drives whose recorded histories gave their respective probabilities of failure as:

$$p_1 = 1/10 \quad \text{and} \quad p_2 = 1/15$$

the probability that *both* would fail at the same time would be:

$$p_1 * p_2 = 1/10 * 1/15 = 1/150.$$

13.8.9 The converse, that both *would not* fail at the same time, is

$$9/10 * 14/15 = 126/150.$$

The sum of these two is clearly *not* 1 so what has 'gone wrong'?

13.8.10 The answer is that there are other possibilities, that one would fail and the other not. This can happen as either

$$9/10 * 1/15 \text{ or } 1/10*14/15.$$

13.8.11 Can you see why? If we now add *all the possibilities* we find

$$\frac{1+126+9+14}{150}$$

which, happily, equals 1.

13.8.12 The probability that **either** one or the other would fail at any given time would be:

$$p_1 + p_2 = 1/10 + 1/15 = 3/30 + 2/30 = 5/30 = 1/6.$$

(The data used in the example are absurdly large failure rates for disk drives but are chosen to be simple enough for the *method* to be made clear without the arithmetic becoming too difficult.)

13.9 Probability trees

13.9.1 If we take a still more complicated example and have *three elements* or more in our system configuration it would make for quite complicated arithmetic.

13.9.2 It is then that it is very useful to turn to another type of model which we have met before.

13.9.3 Suppose, still using much simplified numbers, that we have a CPU, a VDU, a disk drive and a printer, with probabilities of failure of 1/20; 1/15; 1/50 and 1/10 respectively. There will be many possible combinations of these four elements.

13.9.4 To define fully the probability space we may use a binary tree model, but this time turned to the horizontal rather than vertical. In the example shown in Figure 13.7 the branchings, from left to right, are the pieces of equipment in turn. The binary branchings point upwards if the probability is that the item will work and downwards if it will not.

Figure 13.7 The vulgar fraction binary tree.

13.9.5 On each path, following the sequence of binary branchings, is the probability of each of the possible *independent* events and the complete tree embodies the whole of the probability space. By selecting the appropriate sequence of branchings, for any path from start to finish, any combination of probabilities may be read off from the tree.

13.9.6 For example we have, as the probability that all four items will work, the sequence of four consecutive upwards-pointing paths. Since this is an AND sequence we have, as the final outcome,

$$p = 19/20 * 14/15 * 49/50 * 9/10$$

all different vulgar fractions and multiplying them together involves considerable 'number crunching'.

13.9.7 We must have an eye for accumulated error in such circumstances but this is a case where the common base of *decimal fractions* will serve us better and we can state them as

$$p = 0.95 * 0.93 * 0.98 * 90$$

The answer, 0.782, indicates that there is slightly better than a 78% probability of a trouble-free system at any given time, even with such poor reliability as our specimen probabilities indicate.

13.9.8 It is often the case that properly *calculated* probabilities are quite different from what we may have expected so it is wise always to double-check our results. Above all remember that the answer cannot be less than zero or greater than one!

13.9.9 Now take all possible paths through the tree and put the probability of each 'and' outcome as the 'leaf' at the end of each path. What *should* the total of all the leaves be? What total do you arrive at?

13.9.10 From the same tree the probability that, for example, EITHER the processor OR the VDU OR the printer is not working at any given time will be

$$1/20 + 1/15 + 1/10 = 13/60 \text{ or } 21^2/3\%.$$

This assessment is made by omitting the disc drive; can you see how this has been done?

13.10 Other experiments

13.10.1 There remains only one other complication for us to explore within this topic and we can see this most clearly in the context of some traditional practical experiments.

13.10.2 If we place three white counters and three black counters into a bag and, whilst blindfolded, remove one counter, what is the probability that it will be white?
It is perhaps not too difficult to see that

$$p = 3/6 = 1/2.$$

If we *keep out* the counter we have taken, which *did* happen to be white, what is the probability that a second 'draw' will also produce a white counter? We have only five counters left, of which two are white so p_2 is 2/5. This shows us that the **contingent** probability of two white counters in succession is 1/2 * 2/5 or 1/5, 20%.

13.10.3 Now perform the experiment slightly differently *by putting the first counter back* before drawing the second. Both **independent** probabilities would now be 3/6 or 1/2 and 1/2 * 1/2 = 1/4 = 25%. There *is* a difference and quite an important one between the two conditions.

13.10.4 In our original test the probability of taking *either* a white *or* a black counter is 1/2 + 1/2. Since this equals one the outcome is certainty. This seems self-evident since *the counter must be one or the other; there is no other possible outcome.*

13.10.5 If we put three white, three red and four black counters into our bag the probability of taking out *either a white or a red ball* will be

$$3/10 + 3/10 = 3/5.$$

13.10.6 The first 'either' outcome is 3/5 and the second is 7/10 (by the same additive process if the first counter was replaced in the bag, but 7/9 if it was not). The two 'eithers' are, however, joined in the question by an 'and' which means that the two 'partial outcomes' must now be multiplied together: 3/5 * 7/10 = 21/50 or 42%. such models.

13.10.7 We can go on to take further steps to elaborate such experiments almost indefinitely with questions like, 'what is the probability of finding either a red or a white counter on the first trial and then either a red or a black counter on the second trial?'. With each additional complication the arithmetic becomes a little more involved and we may need to practise each step thoroughly until we are more familiar with it.

13.10.8 This is another type of situation where, if we need to deal with more complex problems any good probability text will show us the necessary procedures.

13.11 Conclusion

There is, fortunately, much in quite elementary organization of data which allows us to derive probabilities of events without extremely elaborate procedures and some of our existing tools serve us very well in this respect.

Some further theoretical knowledge is, however, necessary in deciding which tools to use in which kinds of situation.

Exercises

(1) From within a computer staff of 20 people, four are to be selected for a project team of whom one is to be the team leader.
  (a) In how many ways could the choice be made?
  (b) If the team were selected at random what is the chance that one particular person will be in the team?
  (c) If a brother and sister work in the department what is the likelihood that:
    (i)   both will be selected and
    (ii) one of them will be in charge?
  (d) If the first selection were made and then cancelled so that a further selection had to be made on the next day, each choice at random, what is the probability that brother and sister will each be chosen once only, but not on the same day?

(2) Five of a computer staff of 22 are programmers.
  (a) If you walk in to the department what is the probability that the first person you met will be a programmer?
  (b) Does your answer to (a) change and if so how, if you are a programmer in that department?

(3) From the data in example (2) from the Exercises for Chapter 12:
  (a) what is the probability that any individual person will be:
    (i)   not older than 40;
    (ii) more than 80 years old;
    (iii) between 50 and 60 years old?
  (b) if any three were selected at random what is the probability that none of them would be over 60 years old?

(4) In a set of examinations 60 candidates gained marks which showed the following statistics:

	Mean	SD
Maths	48	16
Programming	55	12

  (a) If one candidate scored 80 for maths and 65 for programming what were his or her 'standard scores'?
  (b) The percentage of candidates normally scoring more than given standard (z) scores is shown in this table:

z score	0.4	0.6	0.8	1.0	1.2	1.4	1.6	1.8	2.0	2.2	2.4
Percentage scoring higher	34.5	27.4	21.2	15.9	11.5	8.1	5.5	3.6	2.3	1.4	0.8

(i) What is the probability of each of the scores in **(a)**?
(ii) How many people out of those sitting might be expected to have done better in each subject?
(iii) What is the probability of anyone gaining this combination of marks?

14

Applied two-state logic and switching

Objectives

After working through this chapter you should be able to:

❑ understand the importance of two-state logic in computing
❑ model logical relations by truth-tables, Venn diagrams, switching circuits or 'gates'
❑ analyse and simplify circuits by Boolean algebra
❑ use Karnaugh maps.

14.1 Logic in programming

14.1.1 Sooner or later, in all but the simplest of programs, we meet a situation where we have to write something like, 'IF *so-and-so* THEN DO *something*, ELSE DO *something different*'.

This means that we *must* be able to model the problem in two-state logical terms, where the circumstances are so closely defined that the 'IF' can only be answered by yes or no; there may be no doubt.

Such absolute statements are quite foreign to ordinary discourse between humans which is full of approximations and shades of meaning.

14.1.2 It is for these reasons that failures of *logic* within our programs are just as common as failures of the syntax of the language we happen to be using. Logical failures often far more difficult to detect and eliminate. Indeed if our logic *is* in any way defective it may not become apparent until long after the program has gone into use when some unusual set of circumstances or piece of data, which we had not foreseen or tested for, just happens to come along.

14.1.3 The *logic* of a program is entirely *independent* of the programming language used and we should be able to construct our models in pseudo-code and test our solutions properly, before final coding, to make sure that they do precisely what is intended.

The situation is often complicated by the fact that we may be making several logical decisions within a single line of code. If, for example, we are sorting data into the classes which we used for our statistical distribution of the run-times, early in Chapter 11, we might well do so by a line such as:

IF $x < 7$ THEN $c_1 \leftarrow c_1 + 1$ ELSE IF $x < 11$ THEN $c_2 \leftarrow c_2 + 1$ ELSE IF $x < 15$ THEN $c_3 \leftarrow c_3 + 1$ ELSE IF $x < 19$ THEN $c_4 \leftarrow c_4 + 1$ ELSE IF $x < 23$ THEN $c_5 \leftarrow c_5 + 1$ ELSE IF $x < 27$ THEN $c_6 \leftarrow c_6 + 1$ ELSE $c_7 \leftarrow c_7 + 1$.

14.1.4 This involves **compound** decisions within which *each single decision* must be *logically* correct, without any doubt whatever.

14.1.5 This is **rigorous** or **two-state logic**, which we have already met, but in a slightly different kind of application. We refer to it also as **binary logic** and in the textbooks we use '1' for 'true' or 'yes' and '0' for 'false' or 'no'. For technical reasons, which you may be able to relate to the work done in Chapter 6, many processors use '−1' instead of the '0' for **negation**, as the 'no' is often called.

For simplicity we will follow the '1/0' convention in this book.

14.2 Logic and hardware

14.2.1 Apart from this fundamental importance in programming we also need to know precisely how the *logical* two-state system maps on to a *physical* two-state switching system if we become involved with assembly languages or the theory and architecture of the computer, or if we need to program any process control applications.

14.2.2 Today such systems, from controlling traffic lights to running the cycles of a washing machine, embody the logical models in various arrange-ments of miniaturized transistor switches in 'chips'. In this context 'yes' ordinarily translates as 'switch *on*' and 'no' as 'switch *off*'.

14.3 Language and symbols

14.3.1 We saw in our chapter on **sets** how the strict mathematical conditions which govern them allow us to define *all* the relations in a given universe of discourse in a Venn diagram. We shall make further use of such diagrams in this chapter but in spite of their very great usefulness they do have limitations.

Try drawing a Venn diagram with four or five intersecting sets where each pair of sets has an intersection and where there is a subset embracing *all* of them. You will, I think, see how much more difficult it becomes.

The Venn technique is also much more powerful for evaluating *enumerated* sets than it is for solving more complex problems about the very basic relationships *between* sets.

14.3.2 We shall, therefore, need to investigate the two-state logical world a little more deeply and in such further study there is, as always, some different language and some new symbols to be learnt.

14.3.3 Among the jargon *words* we meet **statement** in a new usage. It is applied *very strictly* in two-state logic to mean any *single* proposition which can *only* be identified as either **true** or **false**.

'He is male' is a statement which satisfies these new criteria, 'He is a man' is *not*. This is because the set of males *is* defined but the age at which a boy becomes a man is not. The second sentence involves what is *arguable* and it is not, therefore, a *statement* within our strict definition.

14.3.4 Unlike the material of the previous chapter the study which we now enter admits *only* of **total certainty**. That certainty associates 'yes' with 'true' and binary 1; 'no' with 'false' and binary 0.

14.3.5 Statements may be **compound**, when we associate two or more by the *conjunction* AND, or by the *disjunction* OR. 'He is male and he is a programmer', is a **conjoint** compound of two statements. It is only *ultimately* true if *both parts* are true.

'He is a policeman or a thief' is a **disjoint** compound statement which is ultimately true if *either* OR *both* part(s) is/are true. It is worth looking at these two examples more carefully so as to get the foundations absolutely secure. Since, as we saw in the chapter on sets, every IS has a corresponding IS NOT the first tells us that a male may be *not* a programmer and that a programmer may be *not* a male. The *compound* statement is true, however, only if the 'he' in question IS both.

The second compound tells us that 'he' may be *not* a policeman or *not* a thief, but the compound statement is true if 'he' is *either.*

14.3.6 We often use the small letters *p, q* and *r* as the symbols for this sort of logical statement and we clearly need symbols for 'not', as well as 'and' and 'or'. In our first example we might allocate '*p*' to 'is male' and '*q*' to 'is program-mer' and we would write the compound statement as '*p* ∧ *q*' to mean 'the consequences' of the two statements *in conjunction*.

I say *consequences* because there are *four* possible results of any compound of two statements, as we shall see a little later.

We would write, of our second compound statement '*p* ∨ *q*' to indicate the possible results of the *disjunction* of *p,q*. The 'not *p*', or 'it is false that he is a policeman', is symbolized by ~*p*.

14.4 Truth tables

14.4.1 By now the ideas and the symbols may have begun to cause some confusion so let us use our primary tool for simplifying such models, the **truth table.**

14.4.2 There are, for any two statements in compound, four possible relations. Any '*p*' may be *only* either true or false and any '*q*', likewise, may have only these two states. This leads to a table such as that shown in Table 14.1.

Table 14.1

p	*q*	*p* ∧ *q*
True	True	True
True	False	False
False	True	False
False	False	False

The *disjunction*, on the other hand, gives Table 14.2.

Table 14.2

p	*q*	*p* ∨ *q*
True	True	True
True	False	True
False	True	True
False	False	False

14.4.3 Each statement has its own **truth value** and the outcome of the compound statements is dictated by the ways in which the individual truth values are associated. The tables contain *all possible results* of such associations.

We can see the compound relations, at this very elementary level, in Venn diagrams which, if we look a little more carefully, reveal the need for a *third* compound. This is shown in Figure 14.1.

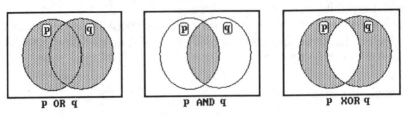

Figure 14.1   The OR and XOR domains.

14.4.4 When we use the word 'or' in everyday life it is often ambiguous. 'These seats are reserved for elderly or disabled persons' reads a notice in a bus: 'Admission half-price for Children or Senior Citizens' reads a poster on the side of the same bus. Can you see the important difference between the uses of the word 'or'? If you are in any doubt look again at the Venn diagram.

The first statement permits of people being *either* elderly *or* disabled *or both*, whilst the two groups in the second statement are *mutually exclusive*; no-one can be *both*.

14.4.5 In logic we distinguish between OR, meaning in either of the named sets *or* their intersection, XOR (exclusive or), which means in either *but not both*. (There *is* a symbol for XOR, $\veebar$, but it is not universally accepted and is less reliable than the others.)

The truth table for XOR is shown in Table 14.3.

Table 14.3

p	q	p XOR q
1	1	0
1	0	1
0	1	1
0	0	0

14.4.6 All such statements, in programming, are set in an IF ... THEN program line but the word 'if' is also ambiguous in ordinary speech and needs clarifying for our purposes.

In some programming languages we find both IF and the IFF which has been invented to represent **if and only if** in order to eliminate the ambiguity. The former describes a situation, symbolized by '$p \rightarrow q$', which says 'if p is true then *q* is also true'. Or, the truth of *q* **depends** upon first *p* being true and the *sequence* is critical.

14.4.7 Iff, on the other hand, goes much further and, symbolized in turn by '*p* ↔ *q*', says '*p* is true *if and only if q* is also true'. They are **mutually dependent;** neither can be true without the other also being true.

For example, 'a customer number is valid **if** the check-digit routine sums to zero', certainly does *not* imply that if the checksum is zero the number is valid; there will always be cases, although low in probability, where the checksum will still be zero if there has been a peculiar transposition of digits. Yet we quite commonly find even experienced programmers and analysts making such assumptions, but it is merely an **if** not an **iff.** 'The customer may have a 5% discount **if** his account remains within his credit limit', is fundamentally different. It is an **iff.** If he is within his credit limit he may have the 5% discount but if he receives a 5% discount he *must* be within his credit limit. Programming and testing such differences is not always easy.

14.5 Logical equivalence

14.5.1 The use of truth tables leads also to another condition, the **logical equivalence** of compound statements. If the outcomes of two truth tables are identical the statements which they represent are *logically equivalent* and we use the '≡' symbol to represent this state.

14.5.2 For example, let us look at: ~(*p* ∧ *q*); and (~*p* ∨ ~*q*). Since the 'outcome' columns of each section of the table are exactly identical we can safely say that: ~(*p* ∧ *q*) ≡ (~*p* ∨ ~*q*). We commonly present such tables as a direct comparison as shown in Table 14.4.

14.5.3 Such programming constructs as we have been discussing are known as **conditionals** and it is important, in many programming applications, that we should test them rigorously by verifying their truth tables.

Table 14.4

p	*q*	*p* ∧ *q*	~(*p* ∧ *q*)	*p*	*q*	~*p*	~*q*	~*p* ∨ ~*q*
1	1	1	0	1	1	0	0	0
1	0	0	1	1	0	0	1	1
0	1	0	1	0	1	1	0	1
0	0	0	1	0	0	1	1	1

14.5.4 The compound statements will often include more than two propositions and the table may then become quite large. For example 'if p and q and r then ...' will generate Table 14.5 below in which, for each of the *four* possible results of combining p and q there will be *two* possible conditions of r giving Table 14.5.

14.5.5 If the compound statement were 'p and q or r' we might find that the interpreter we are using *automatically* takes the 'and' before the 'or' but it is often safer to use brackets whilst we are working on the model and write it as $(p \wedge q) \vee r$ and to insert an extra column into our table to accommodate the intermediate result, as shown in Table 14.6.

Once again we see all possibilities and our testing of such a model, in program form, must include all the combinations which can give rise to the eight possible results.

Table 14.5

p	q	r	$p \wedge q \wedge r$
1	1	1	1
1	1	0	0
1	0	1	0
1	0	0	0
0	1	1	0
0	1	0	0
0	0	1	0
0	0	0	0

Table 14.6

p	q	$(p \wedge q)$	r	$(p \wedge q) \vee r$
1	1	1	1	1
1	1	1	0	1
1	0	0	1	1
1	0	0	0	0
0	1	0	1	1
0	1	0	0	0
0	0	0	1	1
0	0	0	0	0

14.6 Switching diagrams

14.6.1 It has long been the practice to illustrate such possibilities, in **control systems**, by reference to diagrams showing the switching **on** and **off** of an electric light, as shown in Figure 14.2. If you were not very good at science at school don't worry, just try following the track of the diagram with a pencil. If you can get from one 'supply' point to the other without meeting any gaps the condition is '**on**' and the lamp would light. Otherwise it will not.

14.6.2 The switching function is carried out, within the computer, by circuits known as **logic gates**, for which logical statements have to be converted to electrical voltages, still identified as '1' or '0' for 'on' and 'off' but with individual *statements*, or **inputs**, labelled as *A, B, C,* and so on, whilst *consequences*, or **outputs**, are labelled *X, Y, Z*.

Figure 14.2 Three switching circuits.

14.6.3 These 'gates' are represented in diagrams as in Figure 14.3. but, just as with a truth table, compound statements may give any number of 'inputs' and gates may be combined in many different ways.

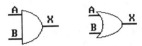

Figure 14.3   The two principal logic gates.

14.6.4 The symbols are also different. In this field the 'and' conjunction of statements, (or the intersection of A and B) is shown as **A.B**.

14.6.5 The dot *on the line between the two 'variables'* is a long-standing multiplication sign. The alternative is simply *AB*, as in ordinary algebra but here meaning, not *A times B* but *'A and B'*. Where we write *A + B*, on the other hand, we mean *'A or B'*; a very common cause of confusion!

14.6.6 This leads to a variation in the writing of the truth table such as shown in Table 14.7.

Table 14.7

A	B	X = (A.B)
1	1	1
1	0	0
0	1	0
0	0	0

This represents all the possible conditions for the two-input AND GATE symbolized in Figure 14.4, and means that to generate a voltage, indicated by 'one', at the *output*, we must apply voltages at *all inputs*.

Figure 14.4   The AND gate.

14.6.7 The *disjunction,* on the other hand, uses the 'OR' gate and is represented by its own different version of the truth table, still with *A, B* and *X,* but using the '+' symbol for the 'OR' relationship.

Figure 14.5 An OR gate.

In this case a voltage, (1), at the output is generated by an appropriate voltage at *any one or more inputs*. The table is shown in Table 14.8.

Table 14.8

A	B	X = (A + B)
1	1	1
1	0	1
0	1	1
0	0	0

Each table, once again, contains all the possibilities about the gate concerned.

14.6.8 The XOR gate is shown in Figure 14.6 for comparison.

Figure 14.6 The XOR gate.

14.6.9 If we wish to look at *combinations* of gates, as they represent more elaborate sequences of statements, we can start with the equivalent of the third of the light switching circuits in Figure 14.2, earlier.

In Figure 14.7 we see that our original *p* and *q* are represented by *A* and *B* and are inputs to an AND gate. The output of this is one of the inputs to an OR gate whose other input, *C* represents the *r* of our earlier model of the same situation. We see here that *either* (*A and B*) or *C* will cause an output of one at *Y* in the diagram.

Figure 14.7 AND combined with OR.

As a truth table this will closely resemble the earlier version but once again the symbolism will have changed; see Table 14.9.

Such models are the very essence of design strategies for complex switching applications of computers and are often 'hard-wired logic' systems.

Table 14.9

A	B	(A.B)	C	(A.B)+C
1	1	1	1	1
1	1	1	0	1
1	0	0	1	1
1	0	0	0	0
0	1	0	1	1
0	1	0	0	0
0	0	0	1	1
0	0	0	0	0

14.6.10 There is no limit to the complexity of switching circuits but we do, quite often, need to **invert** an input or an output to conform to our logical model.

Suppose, for example, that we have as a statement, 'if in A and *not in B* then DO ...'. Somehow we have to be able to input 'NOT B', commonly written as $\overline{B}$, and the 'gate' used for this purpose is symbolized in Figure 14. 8, where we see the A input transformed into the 'not A' or $\overline{A}$, also written, you may recall, as A'.

Figure 14.8   Invert gate.

14.6.11 When we follow this a little further we may see circuits of the sort of complexity of Figure 14.9 where the combination of two 'inverts' three 'ands' and one 'or' may, at first sight seem quite bewildering.

14.7 Analysing circuits

14.7.1 One way of trying to understand better what is happening is to label *all* the paths through the diagram in Figure 14.10.

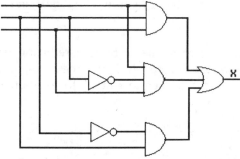

Figure 14.9 A complex of gates.

Figure 14.10 The diagram when labelled.

14.7.2 If we follow each input, through any 'inverters' if present, to each of the gates in turn we can see that each output from the corresponding 'and' gate can then also be written.

By definition each of these outputs will be a 'one' or a 'zero' and *they* serve as the inputs to the 'or' gate which, as we saw earlier, is represented by the '+' sign.

14.7.3 Reading the outputs from the 'and' gates, from the top down, and combining them by the 'or' gate we see the final output to be a combination of all the individual inputs to the final stage.

14.7.4 This is not, however, the end of the story since such switching circuits are often over-complicated because they *duplicate*, or even triplicate or quadruplicate, functions.

Often this is not immediately apparent, even to the designer. It is sometimes even less obvious to *us*, when we are putting compound logic statements into our programs. The results of rigorous simplification can, however, be quite startling.

14.7.5 In Figure 14.11 we see the result of reducing to a minimum the gates needed to provide the exact equivalent of the circuit in Figure 14.10 when the rules of strict logic are applied to simplify the relations as far as possible.

Figure 14.11   Figure 14.10 simplified.

14.8 The algebra of logic

14.8.1 At this stage we must investigate the mathematical toolkit which allows us to achieve such results.

14.8.2 If we look carefully at the **notation** used to describe what is happening in Figure 14.10 we see that we are making *complete statements* whilst using *only symbols*.

But, as we saw in Chapter 10, this is precisely what *algebra* is and there is a special algebra, developed originally by a mathematician named George Boole, which precisely fits our needs.

14.8.3 This **Boolean algebra** uses much the same processes as ordinary algebra, especially in *simplification*. It has its own specialized laws as well as slight variations of those with which we have already become familiar. The symbolism is actually a good deal simpler since no coefficients of variables are involved but we do need to remind ourselves that the symbolism of sets and two-state logic include: A; A';$\bar{A}$, for example, and two operators, '+', as in $A+B$, for 'or' and, as in $A.B$, the '.' stands for 'and'.

14.8.4 Using these symbols our original laws, largely the same as those in Chapter 1, may be rewritten as:

(1a) $A+\bar{A} = 1$ (1b) $A.\bar{A} = 0$
(2a) $A+0 = A$ (2b) $A.1 = A$
(3a) $A.0 = 0$ (3b) $A+1 = 1$

14.8.5 These first six may be explored using a simple Venn diagram on the understanding that 'one' represents the whole universe of discourse and 'zero', in turn, represents the **null** condition.

Figure 14.12 One set and the laws.

14.8.6 The next pair of laws relate to a two-input, or two-set, compound statement and may, in similar fashion, be explored using a Venn diagram such as that in Figure 14.13.

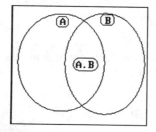

Figure 14.13 Two input diagram.

The two laws:

(4a) $A+B = B+A$ and **(4b)** $A.B = B.A$

do, however, govern two quite different conditions, the 'OR', the UNION, $A+B$, in Figure 14.14 on the next page, together with NOT$(A+B)$.

14.8.7 There is, in law **(4b)** the *intersection* of A and B and Figure 14.15 overleaf shows both this and its 'not'.

14.8.8 The remaining laws, although shown here as three-variable, three-statement, three-set or three-input problems, are all equally applicable to multiple variable models. In this book, however, we shall go no further than

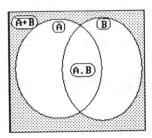

Figure 14.14   *A + B* and its NOT.

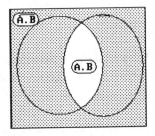

Figure 14.15   The AND with its NOT.

three-component combinations, two of which are illustrated in Venn diagram form in Figures 14.16 and 14.17.

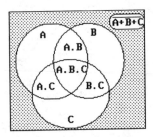

Figure 14.16   *A + B + C* and its NOT.

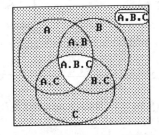

Figure 14.17   *A.B.C.* and its NOT.

14.8.9 The laws, from **(5)** onwards are:

(5a) $A+(B+C) = (A+B)+C$
(5b) $A.(B.C) = (A.B).C$
(6a) $A.(B+C) = (A.B)+(A.C)$
(6b) $A+(B.C) = (A+B).(A+C)^{\dagger}$

Those marked with the dagger (†) are additional to our original set of laws as are:

(7a)	$A.A = A$
(7b)	$A+A = A$
(8a)	$A.(A+B) = A$
(8b)	$A+(A.B) = A$

with one other addition law

(9) $\quad\quad\quad (\overline{\overline{A}}) = A$

These laws govern our use of the tools by which we state and simplify the Boolean algebra of complex logic systems, much of which can be achieved using the algebraic methods we considered earlier.

14.8.10 Look back at our example in Figures 14.10 and 14.11, above, and let us use the laws, step-by-step to simplify what we had originally:

$$X = A.B.C + A.\overline{B}.C + \overline{A}.B \text{ which may be simplified as:}$$
$$X = A.C.(B) + A.C.(\overline{B}) + \overline{A}.B \quad \text{(law (5b))}$$
$$X = A.C.(B+\overline{B}) + \overline{A}.B \quad\quad\quad \text{(law (6a))}$$
$$X = A.C.1 + \overline{A}.B \quad\quad\quad\quad\quad \text{(law (1a))}$$
$$X = A.C + \overline{A}.B$$

which corresponds to what we saw in Figure 14.11.

14.9 De Morgan's Laws

14.9.1 Usually referred to separately are two further laws, known as De Morgan's:

(10a) $\overline{(A+B)} = \overline{A}.\overline{B}$ **(10b)** $\overline{(A.B)} + \overline{A}+\overline{B}$

If, in addition to our 'ordinary' Boolean algebra, De Morgan's laws are used in conjunction with the others in any simplification process, we must follow a strict sequence:

(1) Convert every AND to OR and every OR to AND.
(2) **Invert** all individual 'variables', so that, for example A becomes $\overline{A}$ and $\overline{B}$ becomes just B.
(3) **Invert** the whole final model.

14.9.2 As always the best way to grasp what is happening is to take an example and follow it through the process and then to get plenty of practice with a number of examples.

Let us look at the statement

$$\overline{A.B} + \overline{(A.B)} \equiv \overline{A}.\overline{B}$$

At first sight this seems very odd since the first part on the right-hand side appears to unbalance the equation.

Strict logic is, however, not always immediately obvious to the inexperienced eye and we need to go through the procedure quite deliberately in order to show whether such a statement is true.

Applying De Morgan's laws, we have, after step 1:

$$A+B . \overline{(A+B)} \equiv \overline{(A+B)}$$

and after step 2 this becomes:

$$A+B . (\overline{A}+\overline{B}) \equiv (\overline{\overline{A}.\overline{B}})$$

and, finally

$$(A+B) . (A+B) \equiv A.B$$

14.9.3 The problem may be solved also by comparing truth tables. The whole statement $\overline{A.B} + \overline{A.B} \equiv \overline{A}.\overline{B}$, may be looked at as a left-hand side, which produces Table 14.10.

Table 14.10

A	B	$\overline{A}$	$\overline{B}$	$\overline{A}.\overline{B}$	A.B	$\overline{A.B}$	$\overline{A.B} + \overline{A.B}$
1	1	0	0	0	1	0	0
1	0	0	1	0	0	1	1
0	1	1	0	0	0	1	1
0	0	1	1	1	0	1	1

However, right-hand side produces Table 14.11.

Since the right-hand column of each table is identical the *sequence* of statements must generate identical results and the equivalence is demonstrated.

Table 14.11

A	B	A.B	$\overline{A}.\overline{B}$
1	1	1	0
1	0	0	1
0	1	0	1
0	0	0	1

14.10 Karnaugh maps

14.10.1 There is yet another much-used tool for simplifying complex expressions in Boolean algebra or complex switching circuits, especially if they contain numerous 'OR' gates (disjunctions).

Once again a mathematician is remembered in the name of this device, the Karnaugh map. Rather like Venn diagrams these maps allow us to show all possible combinations of up to six inputs, or variables.

14.10.2 The simplest, for two inputs, is rather like a cross between a matrix and a Venn diagram. It is organised in rows and columns as is a matrix but each 'cell', rather than being a place for storing a variable, is a kind of 'truth domain' representing the combinations of the As and Bs.

In Figure 14.18 we see the 'framework' for such a map which has been filled with the relationships corresponding to each domain in turn. These are ordinarily left blank and if, in a compound statement, that particular relation is present, a tick is put into the corresponding square of the map.

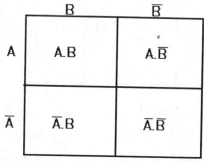

Figure 14.18 The map's domains.

The other preliminary step is to learn the domains of the map and two of the four are illustrated in Figures 14.19 and 14.20.

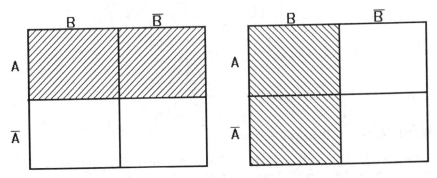

Figure 14.19   The *A* domain.   Figure 14.20   The *B* domain.

14.10.3 Let us now look at the process of simplifying a compound statement such as: $A.B + \overline{A}.B + \overline{A}.\overline{B}$.

We first tick the AB square, next we tick the $\overline{A}.B$ square and, finally, the $\overline{A}.\overline{B}$ square. Next we put a loop around each pair of squares, containing 'ticks' (or crosses for that matter), which touch each other along any one side.

As we saw in Figures 14.19 and 14.20 each pair of such squares is one domain of the map; in our example we have 'looped' the $\overline{A}$ and the B domains and our complete expression can be written as $\overline{A} + B$, in Figure 14.21.

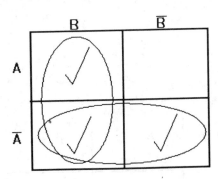

Figure 14.21   The expression 'mapped'.

14.10.4 Check, by truth table or by 'pure' Boolean algebra that this result is correct.

14.10.5 If we move to three-variable expressions we need a rather more complicated map which is really a cylinder, of all things!

This is a rather more complicated idea but is one which we must remember since such maps are normally drawn as two-dimensional but, since they are really continuous, have a kind of 'carry round the back' trick to remain true to the continuous model.

The small gap in Figure 14.22 is to remind us that, even if the cylinder is 'split' here, in order to 'flatten it out to two dimensions, any such split is quite artificial.

We have an exactly similar situation when a flat map of our spherical world is spread out on to a two-dimensional piece of paper, this is also a wrap-around model.

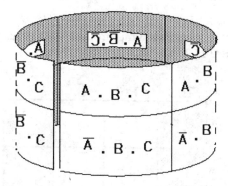

Figure 14.22 The cylinder map.

14.10.6 The only other real complication of the three-variable model is that we 'loop together' not only *pairs* of 'cells' but also any blocks of four and any single 'ticked' squares which touch no others in which there are ticks. Have a look at Figure 14.23.

	B.C	B.C̄	B̄.C̄	B̄.C
A	A.B.C	A.B.C̄	A.B̄.C̄	A.B̄.C
Ā	Ā.B.C	Ā.B.C̄	Ā.B̄.C̄	Ā.B̄.C

Figure 14.23 The tree-variable map.

14.10.7 The domains are different, as we can now see and when we shade the three regions representing *A, B* and *C* we see that all of these consist of *four* 'touching' squares as in Figures 14.24 and 14.25.

Figure 14.24   Domain *A*.   **Figure 14.25**   Domain *B*.

The *C* domain follows the wrap-around pattern explained above and appears as in Figure 14.26.

Figure 14.26   The *C* domain.

14.10.8 Can you identify the domains for NOT *A*, NOT *B* and NOT *C* respectively?

14.10.9 We now have the means of simplifying quite complex statements by assigning each compound of 'ands' to its appropriate square of the map.

Let us take, as a first example, the complex model:

$$A.B.C + A.B.\overline{C} + \overline{A}.B.\overline{C} + \overline{A}.\overline{B}.C$$

The first compound of 'ands', *A.B.C* warrants a tick in the top left square of the map; the second, $A.B.\overline{C}$, needs a tick in row 1, column 2 – and so on.

14.10.10 Once all four terms are represented in the model by ticks (as in

Figure 14.27) we may draw the loops and start the process of extracting the **minimum model**, that is to say the representation of the original *complex* of statements by the fewest possible individual elements.

In our case the first loop, row 1 columns 1 and 2, is wholly in the *A* domain *and* the *B* domain and so is *A.B*. The vertical loop, in column 2, is wholly in the *B* and the NOT *C* domains and is, therefore, *B*. whilst the single circle is in the NOT *A* and NOT *B* and *C* domains and the whole expression is simplified into:

$$A.B + B.\overline{C} + \overline{A}.\overline{B}.C$$

You should, as always, verify this result by one of the other methods we have studied.

Figure 14.27   Three loops.

14.10.11 $A.B.C + A.B.\overline{C} + A.\overline{B}.C + \overline{A}.B.C + \overline{A}.\overline{B}.C$ looks even more complex but produces only two loops, the one shown as two rough semicircles being the complete *C* domain whilst the other is, as in the previous example, *A.B*, which resolves the whole complex of statements into *A.B* + *C*. See Figure 14.28.

Figure 14.28   The wrap-around.

This is the kind of result which so well demonstrates the power of this final tool in the two-state logical world. We need to become familiar enough with these different techniques to be able to use them in our design and programming and to select the one which best suits our purpose at the time.

14.11 Half-adders

14.11.1 As a matter of some practical interest the gates which add numbers in our computers combine into what have come to be known as 'half adders'.

14.11.2 The results which we need are: $0 + 0 = 0; 0 + 1 = 1; 1 + 1 = 0$ carry 1 and, as a block diagram, the circuit is that shown in Figure 14.29.

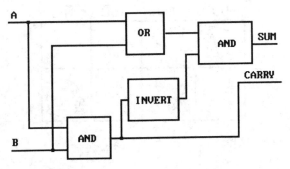

Figure 14.29   The half-adder.

14.11.3 In practice, for technical reasons which need not concern us here, AND and OR gates are much less frequently used than NOT AND, NAND, and NOT OR, NOR, gates which invert the outputs of AND and OR gates, respectively.

14.11.4 The symbols are shown in Figure 14.30.

Figure 14.30   The NOT gates.

14.11.5 Can you work out the truth tables for each of them?

14.12 Conclusion

The world of strict logic makes considerable demands upon our thinking but there are few aspects of computing in which a feeling of competence brings greater satisfaction.

Exercises

(1) Giving the first statement in each case the designation *p*, the second *q* and the outcome *r*, write the truth table for:
   (i)   Kuala Lumpur is in Malaysia or 3 * 7 = 21
   (ii) Kuala Lumpur is in Malaysia or 3 * 7 = 37
   (iii) Kuala Lumpur is in Britain or 3 * 7 = 21
   (iv) Kuala Lumpur is in Britain or 3 * 7 = 37.

(2) Build up the truth table for ~(*p* ∧ ~*q*).

(3) If *p* stands for 'it is windy' and *q*, 'it is raining' write simple sentences to express:
   (a) ~*p*
   (b) *p* ∧ *q*
   (c) *p* ∨ *q*
   (d) *q* ∨ ~*p*
   (e) ~*p* ∧ ~*q*
   (f) ~~*q*

(4) If *p* represents 'Hugo is happy' and *q*, 'Hugo is rich', (assuming for the moment that both are absolute states), write, in symbolic form:
   (a) Hugo is poor but happy
   (b) Hugo is neither rich nor happy
   (c) Hugo is either rich or unhappy
   (d) Hugo is either poor or is both rich and unhappy.

(5) If each of the following pairs of bit sequences were presented to an 'and' gate (as in the illustration), write the sequence of bits which would appear at the output:

   (a) 110001
       101101
   (b) 10001111
       00111100
   (c) 101100111000
       000111001101.

(6) Show the outputs of a 'not' gate which processes the sequences:
   (a) 110001
   (b) 10001111
   (c) 101100111000.

(7) Find a Boolean expression and a truth table for the circuit in the illustration below.

(8) For the following logic circuit:

(a) What inputs will give a 1 at the output?
(b) Write the truth table for the circuit below.

(9) **(a)** Draw the logical equivalent of this switching circuit.
   (b) Give the Boolean expression for the logic circuit.
   (c) By Boolean algebra reduce it to its simplest form.
(10) Draw the Karnaugh maps and reduce to the simplest form the following statements:
   (a) $A = x.y + x.y'$
   (b) $A = x.y + x'.y + x'.y'$
   (c) $A = x.y + x'.y'$
   (d) $A = x.y.z + x.y.z' + x'.y.z' + x'.y'.z$
   (e) $A = x.y.z + x.y.z' + x.y'.z + x'.y'.z$
   (f) $A = x.y.z + x.y.z' + x'.y.z' + x'.y'.z$

(11) For this logic circuit:

(a) Write the Boolean expression.
(b) Draw the Karnaugh map and simplify, if possible.

Answers to the exercises

Many of the questions set will demand that you make a judgement or give weight to the user's needs, as is quite normal in our working situation. Very occasionally the answer may be that there is no solution, as is sometimes the case in the real world. Where you say that this is so it is prudent to give your reasons.

It is rare there is only one way of performing a calculation and there are often several, even many, solutions to particular problems. Within these answers what is given will often be a specimen solution, but where relevant the essential features or steps will be pointed out. You are asked to check your own answer against the one given and to ensure that if it does not follow the same path it arrives at a similar solution or at least contains the minimum elements which are stated.

In the case of models, including pseudo-code versions, names of variables and constants should be stated and the model tested, with appropriate go/no-go data, to ensure that it would serve the purpose.

Chapter 1

(1) Ten animals (domestic/farm/tame are possible classes) – since the individual elements are *unlike* there must be a *common* label.

(2) $5 * 17 * 7 = 5 * 119 = 595$.

(3) The laws which do not apply are: **(1a)** and **(1b)** since $a - b \neq b - a$ and $a/b \neq b/a$; **(2a)** and **(2b)** $(5 - 3) - 2 \neq 5 - (3 - 2)$ and $(6/2)/3 \neq 6/(2/3)$; **(3)**, **(4a)**, **(4b)** and **(6)** (since $n/0$ is not possible in real terms).

(4) $V = 4/3\pi r^3$ – the formula for the volume of a sphere.

(5) **(a) Integer** if stated in years or months only, but **real** if given as years and fractions of years.

(b) Logical, for example 1 = male, 0 = female or **yes/no,** respectively; or possibly **integer** if we use, say, 1 and 2, respectively.

(c) Month – **integer** – $1 \le m \le 12$.

Day – **integer** – $1 \le d \le 31$.

Year – **integer** – $1900 \le y \le 2010$, say. This is a matter of defining a range acceptable to the user and it could be much narrower.

(d) Hour – **integer** – $0 \le h \le 24$ *or* ≤ 12 if 12-hour clock is used.

Minute – **integer** – $0 \le m \le 60$.

Second – decided in consultation with the user according to task. Could be **integer**, $0 \le s \le 60$, if for everyday use but if precision timing is involved, in scientific, process-monitoring or athletic records, for example, it would be **real** and could, in science or computer technology, for instance, involve very small fractions.

(e) Real, positive only, $0 \le price \le$ top of user's price range. **Or** may be validated against a pricing file.

(f) Integer if wholly converted to cents, pence, etc. but **real** if expressed in the normal form $99999.99. For many purposes, especially in the control of error, it is better to use the integer form.

(g) Real in the range $0 \le h \le 168$.

(6) Any two pairs from: (–3,–5); (–2,–3); (–1,–1); (1,3); (2,5).

(7) (–8,–8).

(8) Any three from: complex states; destructive testing; forecasting; opinion testing; reducing uncertainty.

(9) There are many possible answers but any set must pass the tests: **(a)** is it defined? That is to say is there any possible doubt whether an item belongs in or out of the set? If doubt exists the collection is not a set. For general purposes of argument we can accept a less rigorous standard but for strict, two-state logic, there can be no doubt. **(b)** Is each element unique? if any element is replicated the collection may be a 'bag' but it is not a set.

(10) Test as in **(9)** or get someone else to apply the same tests. Above all argue the case for and against each example. Many of the things which we meet in our working experience will not be easily identifiable as sets and we should accustom ourselves to testing by the rules.

(11) Many possible answers, e.g. 'I must not arrive at work later than 8 a.m.', or, 'To eat in this restaurant I must have at least enough money to pay the bill'.

(12) Many possible answers but we should include:

(a) Validate against the variable type, e.g. a **numeric/alpha** user code.

(b) Verify the range.

(c) Validate against a password file.

(d) Check that the code maps on to the password, character by character.

(e) If not valid then 'access denied' message.

Chapter 2

(1) $P = 1.03 * (W + (0.3 * W))$.

(2) Define π, V and r REAL
Accept r
DO UNTIL $r = 0$
$\quad V \leftarrow (4 * \pi * r * r * r)/3$
PRINT r, V
END DO
END

(3) REM let TSV = Total Stock Value; REM open STOCKVALUE file
READ TSV
REM open STOCKFILE
REM open TRANSACTION file
$\quad$ DOUNTIL EOF
$\quad\quad$ READ Stocknumber, Quantity, In/Out
$\quad\quad$ If In/Out = 'OUT' then Quantity $\leftarrow$ Quantity $* -1$
$\quad\quad$ READ from STOCKFILE – key=STOCKNUMBER, Price, Qty. Held
$\quad\quad$ Qty. Held $\leftarrow$ Qty. Held + Quantity
$\quad\quad$ Value = Price $*$ Quantity
$\quad\quad$ TSV $\leftarrow$ TSV + Value
$\quad\quad$ PRINT Stocknumber, Quantity, Value
$\quad\quad$ WRITE to STOCKFILE – key = STOCKNUMBER, Qty. Held
$\quad$ ENDDO
WRITE to STOCKVALUE file, TSV
CLOSE ALL
END

(4) The model must have inserted into it an outer loop of the type:
For $P = 1$ to 6
NEXT P
and must have an array of players' positions subscripted p_1, p_2, and so on.

(5) Between the present lines: WRITE S_i and $i \leftarrow i + 1$ we should need to insert a line such as: if $i > 3$ then $f_i = f_{i-1}/f_{i-1}$.

(6) A model on the lines of: Let d_i be the distance travelled in one month by a lorry then: cost = $((700 + 500) + (2 * d_i))/d_i$ (average \$cost/km for 1 month) and perform this for each vehicle for each month for a year, say, in order to accumulate the data for an overall 'average' cost per km.

(7) (a) Employee's contribution: $T_e = 0.01 *$ gross pay

Employer's contribution: $T_m = 1.5 * T_e$. This can be combined as
Tax = (1.5 * (0.01 * Gross)) + (0.01 * Gross)
= (0.015 * Gross) + (0.01 * Gross)
= (0.015 + 0.01) * Gross – the distributive law
= 0.025 * Gross
or T = (25 * Gross)/1000 if we wish to work in integers as far as possible.

(b) Within the payroll loop, after calculating each individual's Gross Pay and *before* any other deductions are made we should insert something like:
$Tax_i = (25 * G)/1000$
Total Tax Payable ← Total Tax Payable + Tax_i.

(c) Gross Pay must be in the range $0 \leq GP \leq \$5000$, say, or whatever maximum pay may be earned within the user's organization.

(8) Starting record number = S = RND(10), E = S
REM open PAYROLL file
  DOUNTIL EOF
    GET record number E,
    READ GROSS PAY
    If GROSS PAY < 0 OR GROSS PAY > 1000 then DO ERROR
  E ← E+10
  ENDDO
ERROR routine
END ERROR
Close all
END

(9) This is quite demanding and should be taken in stages, taking, say, the A to E table first and then adding the E to A table.

(a) A pseudo-code model should contain something like this: REM Let S = average speed; d_i = the distances between individual stations and T = the time past midnight, 00.00hrs, in minutes, ARR_i = the arrival time at a station and DEP_i = the departure time.
S = 40, T = 420, d_1 = 10, d_2 = 8, d_3 = 14, d_4 = 16
  DOUNTIL T = 1140
  i = 2, DEP_1 = T
    DOUNTIL i = 5
      $ARR_i = DEP_{i-1} + ((d_i/S) * 60)$
      IF i < 5 then $DEP_i = ARR_i + 2$
REM PRINTFILE AtoE
      IF i = 2 then WRITE DEP_1
      WRITE ARR_i
      IF i ≤ 4 the WRITE DEP_i
      i ← i + 1

```
        ENDDO
      i = 4, DEP₅ = T
        DOUNTIL i = 1
          ARRᵢ = DEPᵢ₊₁ + ((dᵢ/S) * 60)
          IF i > 1 the DEPᵢ = ARRᵢ + 2
REM PRINTFILE EtoA
          IF i = 4 then WRITE DEP₅
          WRITE ARRᵢ
          IF i > 1 then WRITE DEPᵢ
          i ← i − 1
        ENDDO
      T ← T + 20
      ENDDO
    Close all
    END
```

This will need to be followed, before printing the tables, by a routine which will convert the times, in minutes past midnight, to the normal hours and minutes form. For each time recorded and INT(T/60) will convert the hours and a T(mod60) routine will give the minutes component. Conversion of the hours to 12- or 24-hour format would then be at the user's discretion.

(b) The dry-run of the program, with the results in tabular form, will show us that the first train in either direction will arrive at its destination only at 08.20. Allowing the turn-around time it cannot start its return journey until the 08.40 service by which time four other trains will have started from the terminus. There will thus be five trains, in all, in each direction before return workings are possible. To provide for the one reserve train a total of 11 will be needed to run the services.

(10) DEFINE *C*, real tariff, *U*, integer
ACCEPT Tariff
ACCEPT U
  IF Tariff = 1 THEN DO ONE
  ELSE IF Tariff = 2 THEN DO TWO
  ELSE DO THREE

  ONE
  C ← 0.28 * U
  DO PRINTROUTINE
  END DO

  TWO
  IF U > 250 THEN C ← 60 + 0.35 * (U − 250) ELSE C ← 60
  DO PRINTROUTINE

```
        END DO

        THREE
        IF U > 400 THEN C ← 120 + 0.25 * (U − 400) ELSE C ← 0.30 * U
        DO PRINTROUTINE
        END DO

        PRINTROUTINE
        WRITE Tariff, U, C
        END DO
    END
```

It must be noted that there are IF ... THEN ... ELSE routines in two of the segments. Make quite sure that you know why and what they achieve.

Chapter 3

(1) **(a)** The set must be defined, i.e. we *must* be able to say of any item that it is either in or out of the set. There can be no 'perhaps'.
   (b) Each element must be unique.
(2) There will be an infinite number of possible answers to this question! The only way to check your own is to ask a friend to apply the two tests in **(1)**.
(3) **(b) (e) (f)** and **(h)** clearly fail the tests in **(1)**. Item **(d)** is a very common problem to database users because it embodies no less than three attributes and all must be the subject of testing questions. 'Is X a customer; does he/she live in Hong Kong; is that a sole residence?' are the questions which expose the problem. The customer who has two addresses or has a business in one place but lives in another often needs very careful definition if the file is not to become a bag. It would be far safer to treat it as doubtful than to say categorically that it is a set as it stands.

(4) a)

b)

c)

(5)

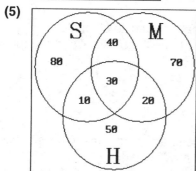

(6) If *T* stands for those who can use all three then 14 can use only one language so the total who can use two or three is 11. So 11 = *T* + (6 − *T*) + (6 − *T*) + (9 − *T*) representing those who can use all three plus each sub-set of those who can use any two *less those who use three.* So 11 = *T* + (21 − 3*T*) = 21 − 2*T* and 10 = 2*T* so the answer is that only five programmers know all three languages.

(7) {1, 8, 27, 64, 125, 216, 343, 512, 729, 1000}.

(8) Using the symbol, → , to signify 'maps onto' :
   (a) $x → x^3$
   (b) $x → x + 32$
   (c) The names of the months in the Gregorian calendar → their respective lengths in days.
   (d) $x → 2x + 2$
   (e) $x → 20 − x$

(9) (a) The set of males over 1.5 m tall
   (b) The set containing all the elements of sets *A*, *B* and *C*.
   (c) The set containing all the elements which are not, at the same time, in all three sets *A*, *B* and *C*.
   (d) The set of elements which are in either *A* or *B* but not both.
   (e) The set of elements which are not in either *A* or *B* or in *A* and *B* combined.
   (f) The set of values from −10 to +10, inclusive, but which are *not* defined as **reals** or **integers** and may, therefore, be either of these.

(10) Any three from:

The database must itself be the *whole* which contains all the parts.

The computer only recognizes two-state logic so the database must be in those terms.

All the information in a database will have relationships with some other parts of the information and relations between sets are the key.

All items in a database must be retrievable and so must have unique attributes.

The database must be ordered and sorted.

Every individual file must be a set.

Records in any one file may be logically 'keyed' to other files.

In short, the database must conform to all the reasons and the rules discussed in this chapter.

Chapter 4

(1) 4.2, –5, 14.9, √2, 0.

(2) Using *N*, *I* and *R* to represent the sets, among the many possibilities are: $N \subset R$; $N \subset I$; $R \not\subset I$; $I \subset R$; $R \neq N$; $R \cap I \neq N$

(3) 8/20 (4/5), 0.4, 40%.

(4) (20 + 15 + 7)/50 = 42/50 = 21/25 or 0.84 or 84%.

(5)

3/16	1/4	5/16	3/8	7/16	1/2	9/16
↓	↓	↓	↓	↓	↓	↓
0.1875	0.25	0.3125	0.375	0.4375	0.5	0.5625
↓	↓	↓	↓	↓	↓	↓
18³/4%	25.0%	31¹/4	37¹/2	43³/4	50%	56¹/4

5/8	11/16	3/4	13/16	7/8	15/16	1
↓	↓	↓	↓	↓	↓	↓
0.625	0.6875	0.75	0.8125	0.875	0.9375	1.0
↓	↓	↓	↓	↓	↓	↓
62¹/2	68³/4	75	81¹/4	87¹/2	93³/4	100%

The differences in decimal values are 0.0625 and in percentages 6¹/4%

(6) Correct to two decimal places.

(7) {√2, √3, √5, √6}

(8) {1, 2, 3, 4, 5, 6, 7, 8, 9}.

(9) The results of squaring {natural numbers}.

(10) In the real world, even with perspective drawing, there are only three physical dimensions. There are *natural* fourth power numbers but we cannot illustrate them.

Chapter 5

(1) 4000 times as large.

(2) 3476.

(3) 30,000,000 – 400,000 = 29,600,000 but, 300 * 10^5 – 4 * 10^5 = 296 * 10^5.

(4) **(a)** 1B7, **(b)** 3E80, **(c)** 7DA0, **(d)** F86.

(5) **(a)** 1199, **(b)** 413, **(c)** 65535, **(d)** 11 259 375.

(6) **(a)** 0.349, **(b)** 285 * 10^2 = 28.500, **(c)** 34.9A Hex, **(d)** 2AF3 Hex.

(7) **(a)** 8, **(b)** 6, **(c)** 4, **(d)** $(10+7+8)_{mod11}$ = 5.

(8) **(a)** 7, **(b)** 11, **(c)** 9, **(d)** 0.

(9) y = 5, 7, 9, 0, 2, 4, 6, 8, 10, 1.

(10) x = 3 is the only element.

Chapter 6

(1) There are many possible sources of mistakes at this stage among which are: document omitted or misread; transposition errors; characters omitted; mis-keying; repeated keys due to keyboard 'dwell' or 'bounce'. Check your answer by exchanging views with a fellow-student perhaps.

(2) Again there are many possible applications such as controlling the completeness of a batch by hashing the document numbers, e.g. invoice numbers, by hashing dates, batching totals of amounts of money, total number of items withdrawn from stores – and so on. Check with a colleague or fellow-student.

(3) Parity checking is testing the make-up of the binary digits in a coded eight-bit byte by making the leftmost digit a 1, if the total number of 1's in the remaining seven bits is *odd* and even parity is being used or, in odd parity, if the total number of 1's is even. For example: 10110111 would show that the byte is correct if *even* parity is being used but it would show that there had been some corruption if the parity were odd.

(4) In each case we multiply each digit in turn by its column weight to give the following table.

Example	Hash total	Mod7 value	Mod7 complement	Final number
(a)	43	1	6	4396
(b)	107	2	5	279465
(c)	18	4	3	300003
(d)	112	0	7	1058537
(e)	208	5	2	174937265

(5) **(a)** Not; **(b)** not; **(c)** not; **(d)** not; **(e)** valid.

(6) No. Only if a pair of digits multiplied by their weighting factors return equal sums.

(7) **(a)** ±0.5 km, **(b)** ±0.05 cm, **(c)** ±0.005 gm, **(d)** ±0.5 cm

(8) **(a)** ±1.4%, **(b)** ±0.37%, **(c)** ±0.038%, **(d)** ±0.0038%

(9) **(a)** The maximum error will be generated if all the top line elements are multiplied together, then all the bottom line elements and finally the division is performed.

(b) (i) Error will be diminished if the sequence is: 3.5/7, 13.5/4.5, 1.5/40.5 giving 1/2; 3/1 and 1/27 =1/18. **(ii)** In the computer we could consider converting all to integers before performing **(b)(i)**.

(10) The *reported* answer would be, on the face of it, 25.375

(a) 3 * 7=21. Relative error against the reported answer −15.8%.

(b) (i) 25.2 if values are cut to 1D: E_r = −0.7%, **(ii)** 25.4 if only the answer is cut: E_r = +0.1%.

(c) 3.5 * 7.3 = 25.55 and E_r = +0.7%.

(d) 25.4 and E_r = +0.1%

(e) Maximum = 25.75525; minimum = 24.99525 (error bound on 2.5 = ±0.05 and on 7.25 = ±0.005). Error bound on the answer is approx ±0.38 and E_r ≈ ±1.6%.

(11) If *r* is not an integer we should, if possible, take its value to 5D and do the same for π. If practicable we should then raise r to the third power, multiply the result by π, multiply by 4 at the next to last step and finally divide by 3. Only by taking such precautions to minimize error could we genuinely claim to be able to state the result to 3D.

Chapter 7

(1) **(a)** 437, **(b)** 001, **(c)** 899, **(d)** 77.

(2) **(a)** (499 + 633) − 1000 leads us to:

$$499+$$
$$\underline{633}$$

and, by 'cutting' the overflow: cut 1 | 132

(b) (499 + 363) − 1000, similarly gives: 862 − 1000 = 138

(c) (2495 + 9208) − 10000 gives: cut 1 | 1703

(d) (1207 + 9350.5) − 10000 cut 1 | 0557.5 = 557.5.

(3) **(a)** 0100 1001 0101

(b) 0010 0010 0010

(c) 0110 0011 0010

(d) 0100 0001 0000.

(4) In BCD addition if the result of adding two digits is a legal BCD code, i.e. 1001 or less then the result stands. If the sum is an *illegal* code, i.e. 1010 or over the *add* 0110 and carry 0001.

(a) 0100 1001 0011 +
0010 0111 0101
0110 10000* 1000 *greater than 1001 so add 0110 and carry 1
0001c 0110
0111 0110 1000 = 768$_{10}$

(b) By the same process the answer will be: 0010 0001 1000

(c) 0001 0100 1001

(d) 0001 0100 0100 0011. Note that there is a carry into the new fourth column.

(5) The point to note in this question is that the 4-bit digits are as they would appear in a partially calculated answer and have not had the 'add 0110 and carry' routine performed. This must be done where necessary to arrive at the appropriate answers: **(a)** 7319, **(b)** 1014, **(c)** 5055, **(d)** 7622.

(6) The key to this question is to remember the 'add 1 to the rightmost place' to convert the one's complement to two's complement form. Since the purpose of using the complement is to *subtract*, all the answers will effectively be negative.

(a) 0 0 1 1 1 0 1 0
1 1 0 0 0 1 0 1 – one's complement
1+
1 1 0 0 0 1 1 0 – two's complement

(b) 11010100, **(c)** 10110111, **(d)** 11011011. Note: a useful check is to convert the two's complement form back into denary and when added to the original number the total should always be 256. Can you see why this should be so?

(7) (a) (i)+32767, **(ii)** -32768

(b) (i) +2 147 483 647, **(ii)** –2 147 483 648.

(8) We are able to instruct the machine to calculate to higher precision and in this way to minimize internal error propagation.

Chapter 8

(1) (a)

(b)

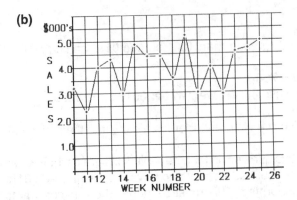

(2) (a) The graph in **(1a)** shows a more rapid growth rate from 2 to 6 years and relatively steady growth therafter. The graph in **(1b)** shows that sales fluctuated considerably from month-to-month.

(b) One might reasonably interpolate in the graph from **(1a)** from 6 to 14 years but there is little practical value in doing so. The extreme fluctuations of graph **(1b)** make it quite useless for interpolation.

(3) (a) The angles would be, approximately, petrol 169.4°, repairs 95.3°, car hire 5.4° and car sales 89.9° but it would be virtually impossible to plot such angles by protractor so that it is sensible to round to the nearest degree and the graph should look something like that below left.

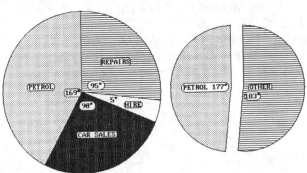

(b) Since the new angle for petrol sales is approximately 177° and only that sector of the graph is specified we might expect the exploded graph to look like that above right.

(c) For a direct comparison we first convert to percentages of the two, three-month period, totals and find that in the first petrol sales accounted for 47.05% of the total and 49.03% of the total for the second period. The increase is, therefore, approximately 2%.

(4) A histogram has a continuous scale as a baseline, normally on the horizontal axis, as a consequence of which the area of each segment is significant. It is truly two-dimensional graph. The block or column chart has a discontinuous scale on one axis and only the other dimension is significant.

(5) (a) Table of moving averages:

Year	3	4	5	6	7	8	9	10	11	12	13	14
Average height	92.5	99.8	105.9	111.4	117.2	122.1	126.6	130.8	135.4	140.2	144.6	149.3

and the graph:

The table for the moving average of sales is:

Week	11	12	13	14	15	16	17	18	19	20	21	22	23	24	25
Average sales	2.75	3.75	3.84	3.42	4.11	4.25	4.33	3.96	4.58	3.79	3.89	3.39	3.99	4.34	4.67

and the corresponding graph:

(b) The trend line inserted into the height/age graph is reasonable if it does not include the first few years where the rate of growth is clearly different. It will be of doubtful value for *extrapolating* for more than a very few years. In the sales graph there could be two quite distinct trends, for the first 20 or so weeks and for the last four, shown at B1 and B2 on the graph. It would perhaps be unwise to take such a short-term view since fluctuations are known to be considerable and the overall trend line, A on the graph, may well be more reliable. The sales staff might prefer to take line B2 as the trend! Mathematically the evidence is not strong for placing great reliance on any of them.

(6) The table of calculated values might look like:

If $x =$	−4	−3	−2	−1	0	1	2	3	4
+4	4	4	4	4	4	4	4	4	4
$2x^2 =$	32	18	8	2	0	2	8	18	32
$2x =$	−8	−6	−4	−2	0	2	4	6	8
then $y =$	28	16	8	4	4	8	16	28	44

and the corresponding graph will appear as below. Note that the graph has been drawn to a very compressed *y* scale for reasons of space. In practice it is always preferable to make the scales as large as the size of paper will permit and in cases such as this, where the *y* values are very much larger than the *x* values, the longer axis should be vertical so as to enhance the accuracy of plotting the graph.

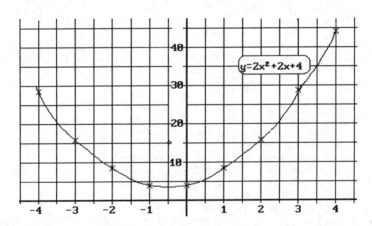

The effect of the changes in scaling may be seen by comparing the figure above with the computer-scaled version on the next page.

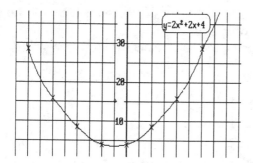

We should always select the scales most carefully and use the largest possible area of graph paper, especially if we may need to interpolate or read off intermediate values from our graph. The other points to remember are to draw a smooth curve, to identify the scales and to give the graph a title.

(7) Once again it pays to tabulate the calculations carefully.

(a)

If $x =$	-2	-1	0	1	2	3	4
$x^3 =$	-8	-1	0	1	8	27	64
$+4$	4	4	4	4	4	4	4
$y = x^3 + 4 =$	-4	3	4	5	12	31	68
$y = 27 - 2x =$	31	29	27	25	23	21	19

and the corresponding graph will be as below.

(b) The point of intersection of the curves on the graph is at $x \approx 2.6$ and $y \approx 22$. If you have drawn your graph to a larger scale and on a larger sheet of graph paper you *may* be able to reach closer approximations. What would you consider the error bound to be in the drawing of your own graph?

(c) The region defined by the constraints will appear as below.

(8) **(a)** If we use L to represent the number of large boxes and S for the number of small each of the functions may be modelled as: sales, $L \leq 280$; $S \leq 400$; production, $480 \geq 1.5L + 0.75S$ since it takes 1.5 minutes to make a large box and 0.75 minute to make a small box. Storage, $4 \geq 0.01L + 0.008S$.

(b) We can then say, for example that if $L = 0$ production may be, at most, $480/0.75 = 640$ small boxes and that if $S = 0$ production may be at most 320 large boxes. In this way we can calculate the three inequalities as:

Sales: $0 \leq L \leq 280$; $0 \leq S \leq 400$
Production: if $L = 0$ then $S \leq 640$; if $S = 0$ then $L \leq 320$
Storage: if $L = 0$ then $S \leq 500$; if $S = 0$ then $L \leq 400$

(c) The graph of the feasibility region will be as shown in below. Within the shaded region, including the lines which make up its boundaries, any variation of large and small boxes is theoretically feasible. The best *practical solution* will, in practice, be decided by the user after taking other demands, such as profitability, into account.

Chapter 9

(1) The plot should look like that below.

1 Square : 50 Knots

(2) (a) The plot of the course sailed will look like that below.

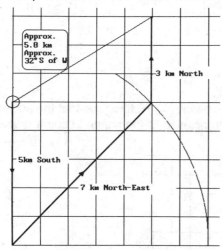

The points to note specially are that the north-east 'leg' must be measured or set out by compasses as here. Similarly the distance remaining and the direction to be sailed must be measured. The alternative description of the course back to the starting point, as an ordered pair or single vector is very much simpler.

(b) The ordered pair is (−5, −3) which represents 5 km west and 3 km south.

(3) (250, −350) plotting the relationships on squared paper will make it clear.

(4) The shape

$$\begin{bmatrix} 1 & 5 & 1 & 2 & 0 \\ -2 & 1 & 1 & 5 & 2 \end{bmatrix}$$

is plotted in the figure. Note that the points are both labelled, A, B, etc., and indicated by a circle. This is done for clarity only and has no special significance except that it allows us to follow what is happening when we transform the shape. In a similar way the scales are numbered on this plot but the numbering will be omitted in later answers for the sake of the clarity of the picture.

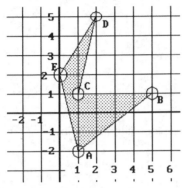

(5) (a) $\begin{bmatrix} 3 & 7 & 3 & 4 & 2 \\ 1 & 4 & 4 & 8 & 5 \end{bmatrix}$

leads us to the figure below in which we see that the addition of the vector leads to the translation of the shape in both the *x* and the *y* axes.

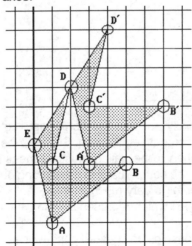

(b) $\begin{bmatrix} -1 & -5 & -1 & -2 & 0 \\ -2 & 1 & 1 & 5 & 2 \end{bmatrix}$

is the next transform which leads to a reflection of the shape as shown in the figure below.

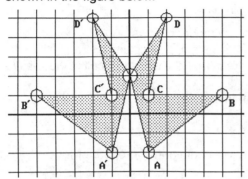

See the figure above. The the transform is a reflection in the *y* axis.

(c) All the values will be multiplied by the scalar, 1.5 to give:

$\begin{bmatrix} 1.5 & 7.5 & 1.5 & 3 & 0 \\ -3 & 1.5 & 1.5 & 7.5 & 3 \end{bmatrix}$

and this in turn will produce the shape below.

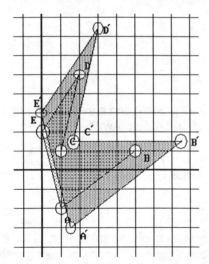

(d) Reflect about the *x* axis.

(e) Move 'up' 2 on the *y* axis and then reflect about both *x* and *y* axes.

(f) Reflect about the *y* axis and stretch.

(6) There will always be many possible ways of programming any matrix operation but one of the most direct ways is by using nested loops with the outer loop, the *i* counter, matching the number of rows and the inner, *j* loop, corresponding to the number of columns. If we designate the 'subject' matrix as *S*, the transform vector or matrix as *T* and the resultant matrix as *R* we can set up some permanent models.

For a *scalar* multiplication:

For i = 1 to r ; rem 'r' = number of rows
   For j = 1 to c; rem 'c' = number of columns
      $r_{i,j} = s_{i,j}$ * Scalar; rem '$r_{i,j}$' = the individual cell of the *resultant* matrix,
  $s_{i,j}$ = the
     individual cell of the subject matrix.
   Next j
Next i

Note: the middle line will contain the '+' operator for a scalar addition.

For a vector addition, adding $\begin{pmatrix} v_1 \\ v_2 \end{pmatrix}$ to the subject matrix, *S* the 'core' line of the model would be: $r_{i,j} = s_{i,j} + v_i$. For a multiplication of, say, a 2 × 5 matrix by a 2 × 2 matrix, as in several of our examples, the core line would be: $r_{i,j} = (s_{1,j} * t_{i,1}) + (s_{2,j} * t_{i,2})$. Any transform involving two or more stages will involve putting together two or more of these operations into the same model.

(7) (a) $A = \begin{bmatrix} -5 & -2 & -1 \\ -1 & 0 & 2 \end{bmatrix}$

(b) $B = \begin{bmatrix} -2 & 4 & 6 \\ -1 & 1 & 5 \end{bmatrix}$, $C = \begin{bmatrix} 1 & 0 & -2 \\ -5 & -2 & -1 \end{bmatrix}$

(c) *B* results in a two times enlargement and a move to the right and *C* results in a rotation through 90° anticlockwise about the origin of the grid.

(8) The addition of the column vector $\begin{pmatrix} -3 \\ 4 \end{pmatrix}$.

Chapter 10

(1) If we use *M* for the matrix representing the machines, *S* for the sub-assemblies, *C* for the component costs, *P* for the prices of assemblies and *T* for the total cost per machine, the operations will be as follows:
(a) With machines A, B and C as the columns and subassemblies 'd' and 'e' as the rows:

$$M = \begin{bmatrix} 3 & 2 & 3 \\ 3 & 4 & 5 \end{bmatrix}, \quad S = \begin{bmatrix} 3 & 2 \\ 4 & 3 \\ 3 & 5 \end{bmatrix}$$

with subasemblies 'd' and 'e' as the columns and 101a, 101b and 101c as the rows whilst the cost matrix, with 101a, 101b and 101c as the columns will be: $C = (3\ 2\ 4)$.

(b) The cost of each machine, in the matrix T, will then be the prices of subassemblies, $P = C * S$, times the make-up of machines matrix, M leading to:

$$(3\ 2\ 4) * \begin{bmatrix} 3 & 2 \\ 4 & 3 \\ 3 & 5 \end{bmatrix} = P = (29\ 30)$$

as the price of assemblies 'd' and 'e', respectively. $P * M$ is then:

$$(29\ 32) * \begin{bmatrix} 3 & 2 & 3 \\ 3 & 4 & 5 \end{bmatrix} = \$(183\ 186\ 247) = T$$

for machines 'A', 'B' and 'C', respectively.

(c) The total cost of one days output will be:

(i) $\begin{pmatrix} 80 \\ 50 \\ 40 \end{pmatrix}$ for the numbers produced multiplied by T

(ii) $(183\ 186\ 247) * \begin{pmatrix} 80 \\ 50 \\ 40 \end{pmatrix} = \$33,820.$

(iii) the new cost vector, $C_2 = (3.15\ 2.15\ 4.16)$ so the new assemblies price will be $P_2 = (30.53\ 33.55)$, $P_2 * M$ will give $T_2 = (192.24\ 195.26\ 262.25)$, and

$T_2 * \begin{pmatrix} 80 \\ 50 \\ 40 \end{pmatrix} = \$35,632.20$ an increase of $1812.20.

(2) In most high-level languages when arrays are declared they must be dedicated to specific variable types. They may hold any legal variable types but not a mixture. In practice we very commonly need to store, in complete records, mixtures of variable types. Linked lists allow us to do this since each list may be associated with a distinct variable type but the association *between lists* is not variable-dependent.

(3) Lists are linked so that records may be held in any required order, or may be ordered as needed on any individual field without the integrity of any record being jeopardized. For example, a company's customer file may, at different demands, be sorted alphabetically by customer name, say, or numerically by customer number, without the records becoming jumbled.

(4) **(i)** A pointer is essentially the continuity indicator of any list, that is to say that it may point to another of the fields which make up the single record, to where the next item in a list is to be found, to where the previous item is held or to the next in sequence of a sorted list. **(ii)** Your answer should be a sketch which resembles that shown in Fig. 10.2.

(5) In a LIFO stack the 'pushing' or 'popping' of a single item does not disturb the rest of the stack whereas in a FIFO stack the whole stack has to be displaced or the complete set of pointers changed when an item is either pushed or popped.

(6) There are many possible answers here and the only sure way to test your own solution is to do a pencil and paper dry run or to write a short program segment and test-run it with suitable data.

(7) Here too there are many possible illustrations. The essential features, however, are that the 'sample' number should always be compared first with the midvalue of the range available, the 'test' value, in our case 75. If the sample number is greater than the test number that the range should be *halved* with the lower half 'discarded'. If, on the other hand, the test value is larger than the sample the upper half should be discarded. If they are equal the search must show a successful end. In our example the first test value will be 75, the second 112 (or 113), the third 94, the fourth 84, the fifth 79, the sixth 77. Only six passes will be needed for our search.

(8) **(a)** Your answer should closely resemble the tree in the figure below.

(b) The table, with its pointers will be:

Node	Contents	Left pointer	Right pointer
1	497	3	2
2	235	7	4
3	502	−1	−1
4	50	5	−1
5	79	6	−1
6	143	−1	−1
7	379	−1	−1

(c) If we assume that the numbers are seen as sorted in *ascending order* the trace and back pointers will be as shown below.

Node	Contents	Left pointer	Right pointer	Back pointer	Trace pointer
1	497	3	2	−1	3
2	235	7	4	1	7
3	502	−1	−1	1	4
4	50	5	−1	2	5
5	79	6	−1	4	6
6	143	−1	−1	5	2
7	379	−1	−1	2	1

(d) 128 would have to fit in as the right pointer from node 6 and the table would be as shown below.

Node	Contents	Left pointer	Right pointer	Back pointer	Trace pointer
1	497	3	2	−1	3
2	235	7	4	1	7
3	502	−1	−1	1	4
4	50	5	−1	2	5
5	79	6	−1	4	8
6	143	−1	8	5	2
7	379	−1	−1	2	1
8	128	−1	−1	6	6

(9) (a) Assuming that the matrices represent three places, *A*, *B* and *C* and that the matrices are in sequence with columns *A*, *B*, *C* and rows *A*,

B and *C* respectively, the networks would be as below. Note: row *A* connected to column *A*, for example, must be shown in some fashion similar to the loops in the figure.

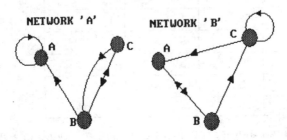

NETWORK ' A'   NETWORK ' B'

(b) The matrices would sum to:

$$\begin{bmatrix} 1 & 1 & 0 \\ 2 & 0 & 2 \\ 1 & 2 & 1 \end{bmatrix}$$

and the network will be as in the figure.

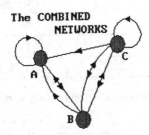

The COMBINED NETWORKS

(c) The links which are one-way only are the sole path from *A* to *B* and one path from *A* to *C*. They are most easily identified if arrows are always put on to the relevant tracks.

(10) (a) If, as in **(9)**, we label the rows from *A* to *D*, downwards, and the columns from *A* to *D*, from the left, only row *D* and column *D* are connected to all of the other stations. The server must, then, be position *D*.

(b) If each station must be linked, by two-way links, to each of the others the relevant matrix will be:

(i)

$$\begin{bmatrix} 0 & 1 & 1 & 1 \\ 1 & 0 & 1 & 1 \\ 1 & 1 & 0 & 1 \\ 1 & 1 & 1 & 0 \end{bmatrix}$$

(ii) we can find the difference between this and what exists at present by subtracting the present matrix from the proposed matrix:

$$\begin{bmatrix} 0 & 1 & 1 & 1 \\ 1 & 0 & 1 & 1 \\ 1 & 1 & 0 & 1 \\ 1 & 1 & 1 & 0 \end{bmatrix} - \begin{bmatrix} 0 & 1 & 0 & 1 \\ 0 & 0 & 1 & 1 \\ 0 & 0 & 0 & 1 \\ 1 & 1 & 1 & 0 \end{bmatrix} = \begin{bmatrix} 0 & 0 & 1 & 0 \\ 1 & 0 & 0 & 0 \\ 1 & 1 & 0 & 0 \\ 0 & 0 & 0 & 0 \end{bmatrix}$$

and this tells us what new paths will be needed.

Chapter 11

(1)

Estimate range	Next estimate	Result†
$1.25 < \sqrt{2} < 1.5$	$\frac{1}{2}(1.25 + 1.5) = 1.375$	$< \sqrt{2}$
$1.375 < \sqrt{2} < 1.5$	$\frac{1}{2}(1.375 + 1.5) = 1.4375$	$> \sqrt{2}$
$1.375 < \sqrt{2} < 1.4375$	do. $= 1.40625$	$< \sqrt{2}$
$1.40625 < \sqrt{2} < 1.4375$	1.421875	$> \sqrt{2}$

and so on to: $1.4140625 < \sqrt{2} < 1.414306641$ and $\sqrt{2} \approx 1.414$ to three decimal places since the difference between the two roots which make the 'estimate range' is too small to affect the rounding to three decimal places. Note that this takes a further *11* iterations and that the 'Result'† is established by *squaring* the next estimate to find if it is greater or less than the root we seek.

(2) Your model may take many forms in detail but it must contain: **(i)** a loop; **(ii)** a test to refuse negative numbers; **(iii)** a test to see if the new estimate is greater or less than the value we seek; **(iv)** a procedure to take the previous *smaller* estimate if the new value is too large or the *larger* previous estimate if the new is too small; **(v)** a 'break-out' routine which will test for an acceptable error bound between each pair of consecutive estimates, for example, 'IF $|e_i - e_{i-1}| < 0.0001$ THEN END'.

(3) (a) The model must accept **real** numbers as input and must reject **negative numbers**. If e_i is any given estimate, $e_1 = \frac{1}{2}x$ and $e_2 = x/e_1$ then: $e_{i+1} = \frac{1}{2}((x/e_i) + e_i)$ and $y = e_i$ when $|e_i - e_{i-1}| <$ acceptable error.

(b) (i) For answers to three decimal places the acceptable error must be ≤ 0.0001. **(ii)** For answers to four sig. figs. we would have to insert a manual 'stop' to the iteration since the range of acceptable x values

is such as to make the number of places in the root quite unpredictable.

(4) At the seventeenth pass we should arrive at: $21.679 < \sqrt{470} < 21.682$
and $\sqrt{470} \approx 21.68$ to two decimal places or after twenty passes we could
write the answer as 21.679 to three decimal places.

(5) You should have found, fairly quickly, that this is a very slow convergence indeed and that it is impractical to perform it by hand. After writing
a short program to perform it, which took several thousand passes, I
found the value of π, to three decimal places, to be 3.142, a difference
of 0.001 or $E_r = 0.03\%$.

(6) $R = (P + T)/H; \ H = (P + T)/R; \ T = (R * H) - R$

(7) (a) $3x^3 + 3x^2 - 2 = y$

(b)

If x =	1	2	3	4
$3x^3$ =	3	24	81	192
$+3x^2$ =	3	12	27	48
-2 =	2	-2	-2	-2
Then y =	4	34	106	238

(8) If $y = 2x^2 + 27$
$y - 27 = 2x^2$
$\frac{1}{2}(y - 27) = x^2$

(9) (a) $6t + 4s = 54; \ 6t + 15s = 153; \ 11s = 99; \ s = 9, \ t = 3.$

(b) The matrix of the left-hand side of the equations becomes:

$$\begin{bmatrix} 3 & 2 \\ 2 & 5 \end{bmatrix} \text{ and its inverse is } \frac{1}{15-4}\begin{bmatrix} 5 & -2 \\ -2 & 3 \end{bmatrix}$$

$$\frac{1}{15-4}\begin{bmatrix} 5 & -2 \\ -2 & 3 \end{bmatrix} * \begin{pmatrix} 27 \\ 51 \end{pmatrix} = \begin{pmatrix} 135-102 \\ -54+153 \end{pmatrix} = \begin{pmatrix} 33 \\ 99 \end{pmatrix} \text{ and } \frac{1}{11}\begin{pmatrix} 33 \\ 99 \end{pmatrix} = \begin{pmatrix} 3 \\ 9 \end{pmatrix}$$

giving the result, $t = 3$, $s = 9$. This appears to be a more laborious
method but, as is explained in the chapter, the matrix method is easy
to program and can work for *any* values whereas the 'traditional'
method depends upon inspection and is quite difficult to program.

(10) $\dfrac{x^5 * x^{-1}}{x^{-2} * x^4} = \dfrac{x^{5-1}}{x^{-2+4}} = \dfrac{x^4}{x^2} = x^{4-2} = x^2.$

Chapter 12

(1) The mean is 25.

(2)(a)

	a_i	f_i
$26 \le 30$	28	5
$31 \le 40$	35.5	109
$41 \le 50$	45.5	168
$51 \le 60$	55.5	188
$61 \le 70$	65.5	105
$71 \le 80$	75.5	14
$81 \le 90$	85.5	1

(i)

(ii)

(iii)

(2) (b) Class

Class	Cu.f	Cu.f (%)
(i) ≤30	5	0.83
≤35	48	8.0
≤40	114	19.0
≤45	205	34.2
≤50	292	48.7
≤55	397	66.2
≤60	480	80.0
≤65	560	93.3
≤70	585	97.5
≤75		
≤80	599	99.8
≤95	600	100

(ii) The ogive should be drawn with as smooth a curve as is possible, as in the figure below.

CUMULATIVE RELATIVE FREQUENCY OGIVE

    (iii) We can also see from the table that the **mode** is 53 and we can read the **median** and the **SIQR** from the graph, below, as approximately 51 and $\frac{1}{2}(58 - 42)$ or approximately 8, respectively.

(c) By the standard procedures the descriptive statistics, taken from the original table, are: **mean** $\approx$ 50.6, **variance** $\approx$ 108.8, SD $\approx$ 10.4. Note that all of these should be stated as 'years' if there is any possibility of their being confused with any other data.

(3) We should calculate and plot the moving 'average' and look to the line of best fit to indicate a trend.

No. The evidence is over too short a period and there could be many possible explanations for such variations. It could be that any one month could be better than average, because of gift-buying on a large scale, say, or similarly that a month could be relatively short, for example February in the Gregorian calendar.

Such a picture could very well be a reason for seeking further evidence but is not, of itself, sufficient to lead to any firm conclusions.

(4)

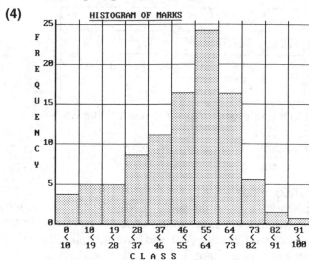

(a) The histogram should look like the one above and should have a title, scales on both axes and no gaps between the 'columns'.

(b) We may derive the approximate mode by inspection and, in this case, since the largest class is clear to see, we can take the mid-point of that class, 59.

In order to calculate the standard deviation and the mean we can profitably use the formula:

$$\sigma = \sqrt{\frac{\Sigma f_i x_i^2}{n} - \left(\frac{\Sigma f_i x_i}{n}\right)^2}$$

but in order to do so we can benefit from making a data table:

Class	X_i	x_i^2	f_i	$f_i x_i$	$f_i x_i^2$
0 < 10	5	25	3	15	75
10 < 19	14	196	5	70	980
19 < 28	23	529	5	115	2645
28 < 37	32	1024	8	256	8192
37 < 46	41	1681	12	492	20172
46 < 55	50	2500	17	850	42500
55 < 64	59	3481	24	1416	83544
64 < 73	68	4624	17	1156	78608
73 < 82	77	5929	6	462	35574
82 < 91	86	7396	2	172	14792
91 < 100	95	9025	1	95	9025
Σ			100	5099	296107
Mean				50.99	

and substituting the relevant values into the formula we have:

$$\sigma = \sqrt{\frac{296,107}{100} - \left(\frac{5099}{100}\right)^2} = \sqrt{2961.07 - 2599.98} = \sqrt{361.12} = 19.00$$

and mean = 51.00, mode = 59.

(c) In this case the range concerned would be from $32 \le x \le 70$ and, by apportioning between the classes at the ends of this range we can arrive at a figure of approximately 69 who have scored in the ±1 SD range. The rule holds good here.

(d) By the same means we can find that *outside 2 SD from the mean,* includes those scoring less than 13 or more than 89 and in our distribution we can find about 7 such candidates, or 7% of the whole. This is nearly double the 4% and this distribution does not conform to that rule.

(5) (a) See the graph below.

(b) The scatter is tightly concentrated around the line marked on the graph and this would indicate that the prices in the two centres *are* closely related to one another.

(6) (a) These data are rather awkward to organise into an odd number of classes with a whole number as the mid-point of each. Seven classes of an interval of $2^{1}/_{2}$ is about the best we can manage and this distribution generates a mean of 67.35 a mode of approximately 67 and a median of 67.8. The SD is approximately 3.1.

(b) There is little to be gained from anything more than a histogram, see below. One difference on this occasion is the 'ruling up' of the modal class in order to find an approximation of the mode.

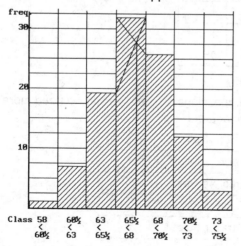

Chapter 13

(1) (a) 4845, **(b)** 1/5, **(c)** 3/190, **(d)** 32/625.
(2) (a) 5/22
    (b) Yes. There are five programmers in all of which you must be one so there are only four others out of a total of 21, if you are the 'onlooker'. The change in probability is small but it *does* exist.
(3) (a) There are 114, out of 600, who are ≤40 so this probability is 114/600 or 19/100 (19%).
    (b) 1/100.
    (c) 188 are between 50 and 60 and roughly 1/10 of those between 40 and 50 might be expected to fit the requirement. Thus approximately 205 out of 600 or 41/140 is the probability we seek.
(4) Standard scores are found by the model (score − mean)/SD so:
    (a) maths $(80 - 48)/16 = 2 = z_m$, programming $(65 - 55)/12 = 0.83 = z_p$.
    (b) (i) 2.3% and approximately 22.1%, respectively (by interpolation), **(ii)** maths 2.3% of 60 = 1 person (since we cannot have fractions!) Programming 13 people. **(iii)** 2.3/100 * 22.1/100 ≈ 50/10000 or 1/200.

Chapter 14

(1) (a) Here the OR is not defined as XOR and we must conclude that the statements will return a '1' if either or both propositions are true. The truth table will be:

p	q	p or q
True	True	1
True	False	1
False	True	1
False	False	0

(2) This involves three propositions, p, q and $\sim q$ and we need more columns to construct the original table:

p	q	$\sim q$	$p \wedge \sim q$	$\sim(p \wedge \sim q)$
True	True	False	False	1
True	False	True	True	0
False	True	False	False	1
False	False	True	False	1

(3) **(i)** It is not windy, **(ii)** it is windy and raining, **(iii)** it is windy or it is raining, **(iv)** it is windy or it is not raining, **(v)** it is not windy and it is not raining, **(vi)** it is not true that it is not raining – it *is* raining.

(4) **(i)** $\sim p \wedge q$; **(ii)** $\sim p \wedge \sim q$; **(iii)** $p \vee \sim q$; **(iv)** $\sim p \vee (p \wedge \sim q)$

(5) An AND gate only returns a '1' at the output when both inputs are '1'. So: **(i)** 100001, **(ii)** 00001100, **(iii)** 000100001000.

(6) **(i)** 001110, **(ii)** 01110000, **(iii)** 010011000111.

(7) $A.B + A.C + A.B.C$ is the Boolean expression since it is an OR gate. The truth table will be:

A	B	C	Y
0	0	0	0
0	0	1	0
0	1	0	0
0	1	1	0
1	0	0	0
1	0	1	1
1	1	0	1
1	1	1	1

(8) $A.\bar{B} + \bar{A}.C$ with the truth table:

A	B	C	Y
0	0	0	0
0	0	1	1
0	1	0	0
0	1	1	0
1	0	0	1
1	0	1	0
1	1	0	0
1	1	1	0

(9) **(a)** The logic diagram, using gates, will be as shown in the figure below:

(b) $A.(B + C) + D.$
(c) This expression cannot be simplified.

(10)

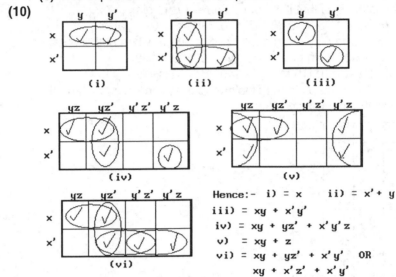

(i)   (ii)   (iii)

(iv)   (v)

(vi)

Hence:- i) $= x$ ii) $= x' + y$
iii) $= xy + x'y'$
iv) $= xy + yz' + x'y'z$
v) $= xy + z$
vi) $= xy + yz' + x'y'$ OR
   $xy + x'z' + x'y'$

(11) (a) By labelling the outputs of each gate in turn we get the following from which we can see that the output from y is any one, two or three of the outputs from the respective and gates.

A.B.$\bar{C}$

B.$\bar{C}$

y

A.B.$\bar{C}$+ B.$\bar{C}$ + $\bar{A}$.B

$\bar{A}$.B

(b) When we put this into a Karnaugh map, as shown below we can see that all the components are contained within the rectangle consisting of: $A.B.\bar{C} + \bar{A}.B.\bar{C} + \bar{A}.B.C.$

B.C   B.$\bar{C}$ $\bar{B}$.$\bar{C}$ $\bar{B}$.C

A

$\bar{A}$

Index

aberrations 212
absolute 116, 215, 240
difference 116, 215
error 98, 99, 102, 144, 206, 249
statistics 19, 230
value 72, 73, 100, 139, 142, 153, 213, 216, 245, 255, 266
address 8, 40, 85, 86, 170, 178
algebra 7, 213, 217–224, 275, 283, 287, 290, 292, 293
algorithms 19
alphanumeric 82, 96, 179
approximation 19, 206, 209, 215, 217, 242, 243, 263
approximations 20, 131, 141, 206, 209, 218, 275
architecture 113, 122, 276
area under a curve 141, 216
arithmetic 19, 32, 57, 68, 76, 81, 82, 87, 90, 91, 96, 97, 104, 107, 114, 115, 118–120, 122, 133, 140, 150, 155, 156, 178, 196, 200, 201, 204, 215, 218, 219, 221, 223, 236, 244, 245, 246, 249, 250, 267, 268, 271
arithmetic mean 245, 246, 250
array 10, 13, 39, 40, 171–173, 177–179, 182–185, 187, 194–196, 199, 201, 228, 229

ASCII 96, 121, 122
assign 8, 31, 109
assigning 7, 8, 295
assignments 31
data 7, 40, 57, 95, 170, 172, 248
associative 3, 223
attribute 14–16, 51, 67, 99
base 76–78, 81–92, 108, 118, 119, 128, 270
number 1–8, 10, 13, 15, 22, 27, 30, 31, 33–37, 39, 40, 42, 43, 47–52, 54, 58, 61, 65–80, 82–84, 86–91, 93, 96–99, 102, 104, 105, 107, 108, 110, 116–118, 128, 129, 133, 134, 144, 145, 160, 161, 165, 168, 172, 173, 179, 180, 182, 183, 187, 193, 194, 196, 204, 207, 209, 215, 217–219, 221–224, 226, 229, 232, 235, 236, 238, 243, 245, 261, 266, 267, 270, 280, 283, 291

BCD 119–121
bicimal fractions 30
binary 30, 81, 83–85, 88, 96, 108, 109, 119, 121, 187–190, 269, 276, 277
binary chop 187
binary coded decimal 119
bits 84, 96, 107, 109, 119, 122
Boolean algebra 275, 287, 290, 292, 293

brackets 31, 32, 61, 158, 219, 221, 224, 249, 281
  sequence 22, 28, 109, 174, 196
byte 84, 96, 119, 178

capture 232
  of data 17
cardinal numbers 68
carry 4, 120, 121, 232, 294, 297
Cartesian 10, 152, 168
cell 40, 161–163, 165, 172, 174, 175, 177, 184–187, 218, 219, 292
characters 7, 95, 96
class 54, 235, 236, 243, 248, 251, 261
class interval 236
codes 7, 96, 119, 121, 185
  numeric 50, 62, 124
coefficients 229, 287
combining 15, 55, 80, 154, 155, 159, 281, 286
commercial model 34
commutative 3, 4, 164, 172
compilers 5, 16, 74, 107, 166
complement 49, 55, 98, 118, 122
complex states 17, 258
congruence 20
constant 7, 30, 34, 139, 142, 166, 172, 195, 226
constructs 21–23, 280
continuous 37, 84, 124, 128, 134–138, 140, 142, 146, 153, 233, 235, 236, 250, 252, 253, 294
  graphs 10, 124, 137, 138, 141
continuous scale 128, 134, 236, 250
convergent 204, 211, 213
cumulative frequency ogive 241, 243, 249, 261

data dictionary 7, 8, 30, 180, 220
data labels 128
data structures 13, 170, 171, 177, 184, 188, 194, 195, 201
data types 7, 30, 34, 96, 179
data-acquisition 29
datum 128, 181–187, 190, 191, 193, 194, 196, 236, 242, 246
De Morgan's laws 290, 291

decimal numbers 30
declaring 7, 8, 30, 47, 62, 75
defined 15, 46, 152, 170, 172, 178, 275, 277
defining 46, 47, 62, 75, 146, 171, 206, 207, 236
denary system 80, 88
denominator 72, 73, 240
Descartes 10
description 2, 6, 12, 29, 33, 47, 48, 54, 99, 244
  narrative 29, 58, 153, 173
destructive testing 231
diminishing uncertainty 230
discourse 16, 47, 48, 52, 53, 55, 140, 260, 275, 277, 288
discrete 128, 129, 134–136, 233, 250, 252
disjunction 277, 278, 283
distortion 126
distribution 197, 235, 236, 239, 241, 244, 245, 251, 261–266, 276
distributive 4, 32, 223
domain 58, 60, 292, 293, 295, 296

EBCDIC 122
element 14, 15, 31, 40, 51, 52, 58, 61, 62, 100, 161, 173, 179, 180, 188, 260
  of set 63, 260
entities 14, 15, 20, 66, 75, 99, 140, 151
enumerating 2
equivalence 2, 20, 21, 145, 146, 218, 223, 227, 280, 291
error 2, 5, 19, 30, 31, 48, 93, 94, 97–110, 114, 122, 124, 140, 144, 145, 206, 207, 211, 219, 224, 235, 247, 249–251, 263, 270
  absolute 101
  induced 99, 122
  inherent 99, 110
  relative 240
error bound 100–103, 105, 145, 206, 207, 211
error propagation 98, 99, 102, 104, 106, 107, 110, 251
extrapolate 137, 147, 215, 217, 252

Fibonacci 41, 43

finite 51, 72, 73, 91–93, 97, 98, 107, 109, 140, 207
fractional numbers 69
frequency distribution 236, 263, 266
function 61–63, 74, 93, 98, 116, 137–140, 142, 144–147, 153, 157, 166, 172, 208, 212, 216, 252, 282
  graph of 146
functions 46, 61, 62, 119, 122, 124, 136–138, 140–142, 144, 146, 147, 159, 166, 178, 184, 209, 214, 215, 229, 231, 286

gates 275, 282–284, 286, 287, 292, 297
  logic 16, 275, 276
generalized model 8
graphs 10, 124–126, 128, 129, 131–134, 136–138, 140, 141, 144, 145, 147, 168, 215, 237, 239, 241, 242
  and spreadsheets 177

hash total 97
heuristic 19, 27, 177, 213
  models 1, 2, 5–7, 9, 10, 14, 17–20, 24, 26, 27, 31, 33, 41, 43, 53, 60, 96, 101, 103, 110, 124, 126, 138, 147, 151, 160, 168, 177, 201, 206, 208, 211, 215, 217, 220, 222, 223, 224, 225, 233, 234, 236, 238, 243, 251, 257, 258, 271, 276, 278, 285, 288
hexadecimal numbers 81, 87
histogram 128, 135, 237–239, 241, 243, 261

identity 5, 139, 159, 165, 167, 175, 176, 180, 228
indentation 23
induced error 94, 99, 103–107, 122, 140, 144, 145, 249
inequalities 20, 21, 124, 145, 146, 218, 229
  and graphs 134
inherent error 99, 100, 102, 103, 144, 207, 247, 249, 263
initialize 8, 174
  variables 8–10, 12, 29, 30, 35, 61, 75, 105, 137, 174, 184, 217, 219, 220, 250, 252–254, 283, 287, 290, 292

integer 30, 52, 70, 85, 99, 122, 171, 178, 179, 219
intercept 139
interpolate 137, 147, 215, 217
interpreters 5, 16, 74, 166
interval scale 134
inverses 4, 5, 115, 223–225, 228
irrational numbers 73
irrationals 73, 109
iterative 34, 142, 204, 206–209, 212, 213, 215

Karnaugh maps 275, 292
key field 15

labels 7, 8, 72, 82, 99, 128, 133, 181, 185, 195, 220
laws 1–6, 19, 86, 90, 115, 123, 139, 156, 172, 222, 223, 287–291
  commutative 3, 4, 164, 172
  distributive 4, 32, 223
  multiplication by zero 5, 176
let 3, 6, 8, 9, 14, 20, 28–31, 34, 36, 37, 39, 41, 46, 47, 51, 54, 56, 57, 59, 66, 70, 79, 86, 90, 98, 99, 102, 103, 107–110, 115, 120, 127, 134, 155, 157, 158, 161, 162, 164, 165, 167, 173, 178, 187, 190, 197, 207–209, 211, 212, 216–218, 221, 224, 229, 233, 234, 245, 246, 248, 260, 268, 278, 280, 290, 291, 293, 295
levels of precision 74
limit 16, 62, 104, 206, 207, 233, 280, 285
linear functions 146
lists 170, 178–181, 185–187, 196
logic 14–16, 21, 47, 48, 51, 56, 60, 188, 274–277, 279, 282, 285–287, 290, 291, 298
logic gates 282
logical 1, 9, 14–16, 23, 27, 46, 47, 50–54, 60, 185, 199–201, 274–277, 280, 282, 285, 297
logical relations 60, 274
logical relationships 14
loop 23, 33–35, 73, 142, 171, 172, 174, 178, 185, 187, 207, 211–213, 215, 217, 293, 294, 296
  conditional 23, 214

loop (*continued*)
   nested 23, 31, 171, 172, 174, 178, 185,
     187
   unconditional 23

machine-readable 21
macros 33, 172
manipulation of sets 16
mapping 61, 62, 73, 74, 77, 78, 157, 172,
   180
mappings 46, 61, 73, 74, 185
mathematical laws 19
mathematical modelling 2, 16
matrices 13, 27, 39, 150, 159, 161,
   163, 164, 168, 171–174, 200, 201,
   228
matrix 10, 13, 39, 40, 124, 150, 156–167,
   171–173, 175–180, 185, 187, 196,
   199–201, 227–229, 292
matrix multiplication 161, 162, 164, 172,
   176
mean 7, 51, 71, 96, 206, 209, 218, 221, 230,
   235, 245–247, 249, 250, 258, 259,
   263–266, 277, 278, 283
   arithmetic 19, 32, 57, 68, 76, 81, 82, 87,
     90, 91, 96, 97, 104, 107, 114, 115,
     118–120, 122, 133, 140, 150, 155, 156,
     178, 196, 200, 201, 204, 215, 218, 219,
     221, 223, 236, 244, 245, 246, 249, 250,
     267, 268, 271
measures of central tendency 235, 243,
   245
measures of dispersion 235
median 241, 245, 249, 262
memory 8, 85, 122, 170, 178, 180–182,
   184, 220, 266
midpoint 236
mistakes 6, 48, 94–99, 114, 177, 204, 224,
   249
mode 1 68, 243, 245, 249
modelling 1, 2, 5–7, 12, 14, 16, 18, 23, 27,
   28, 31, 34, 35, 37, 41, 43, 60, 103, 110,
   126, 144, 147, 152, 156, 174, 177, 188,
   195, 204, 208, 220, 223, 225, 229, 232,
   233, 250, 255, 257
models
   spatial 1, 9, 10, 12, 150–152

statistical 1, 17–19, 96, 101, 133, 134,
   220, 230, 232–234, 248, 255, 258, 265,
   276
symbolic 1, 2, 7, 51, 55, 60, 218, 220
modular arithmetic 76
modulus 91, 98
motion geometry 156
moving average 250–252
multiplication by zero 5, 176

NAND 60, 297
narrative 29, 58, 153, 173
natural numbers 67–70, 75, 82, 99, 208
nested loop 23, 171, 172, 174, 178, 187
nibbles 84, 85, 120
nominal scale 133
non-equivalence 20
non-parametric 14, 18, 133, 233
normal distribution 265
notation 16, 40, 60, 66, 67, 72, 78–80, 82,
   88, 117, 153, 154, 220–223, 229, 236,
   245, 287
number base 76, 77, 108
numbers 14, 39, 68, 69, 83, 99, 105, 118
number system 10, 67, 78, 86, 88–91,
   223
number systems 76, 93
numerals 2, 40, 67, 68, 77–82, 95, 97, 107,
   117, 119, 121
numerator 72, 73
numeric codes 96, 185
numeric functions 62
numerical methods 206, 211
numerical models 41

one-to-one correspondence 62, 63, 77
operators 2, 13, 86, 219, 260, 287
ordered pair 10, 13, 39, 40, 152, 153
ordinal 68, 96, 133, 233
overflow 107, 108

parameter 7, 8, 241, 255
parity 96
pi 7, 61, 62, 132, 164, 206, 207
pictogram 125, 129
pointers 170, 179–183, 185, 190, 192–197
population 56, 107, 234, 264, 265

powers 79–81, 213, 222
  of numbers 222
precision 70, 71, 73, 74, 105, 107, 207, 215
  levels of 241
probability 27, 59, 60, 136, 257–271, 280
  models 19, 41, 177
  trees 27, 170
programming constructs 21, 280
propagation 98, 99, 102–104, 106, 107,
  110, 251
pseudo-code 1, 21–23, 33, 35, 50, 108,
  115, 141, 174, 175, 187, 208, 212, 236,
  276
  modelling 21, 177

quantum 66, 67, 153
quartile deviation 242, 244
quartiles 241, 242
queues 170, 184

radian 166
random 8, 40, 97, 196, 234, 250, 254, 260,
  261
ratio scales 138, 233, 253
rational numbers 72
real numbers 30, 49, 50, 70, 71, 74, 87,
  107
recursive 23, 35, 204, 206, 207, 213
reflections 160, 164
register 107, 109, 117, 120, 122
relations 20, 46, 49, 52–54, 60, 61, 274,
  277–279, 287
  logical 14, 47, 275, 280
relative error 101
reported value 100–103
resultant 155
rotations 165
rounding 104–107, 110, 206, 211, 249

sample 49, 63, 90, 208, 212, 234, 240, 253
sample space 90
sampling 233, 234
scalar 157, 161, 172, 176
scale 10, 14, 70–72, 99–101, 116, 126,
  128–130, 133–137, 140, 142, 144–146,
  155, 157, 174, 213, 222, 235, 236, 241,
  243, 250, 258, 259, 263, 265

interval 133, 236
  number 4, 39, 65, 72, 84, 97, 128, 129
  ordinal 133, 233
  scales 124, 127, 133, 134, 138, 140, 144,
    173, 233, 242, 253
scientific notation 80
selection 22, 33
sequence 3, 4, 13, 22, 27, 28, 31, 33, 36,
  37, 39, 43, 62, 67, 68, 96, 97, 105, 108,
  109, 118, 133, 144, 153, 154, 160, 161,
  172, 174, 186, 187, 190, 191, 193, 196,
  207, 208, 213, 219, 250, 258, 266, 269,
  280, 290, 291
set 2–4, 7, 10, 14–17, 21, 30, 33, 34, 39, 40,
  43, 46–58, 61–63, 66–70, 73–75, 77,
  91, 96, 103, 107, 108, 116, 117, 122,
  140, 142, 144–146, 156, 160, 165, 167,
  170, 171, 173, 180, 182, 185, 190, 191,
  196, 198, 208, 209, 213, 215, 217, 226,
  227, 236, 244, 245, 247–249, 252,
  259–261, 267, 275, 277, 279, 288, 290
significant figures 74, 106, 107, 110, 206
simultaneous equations 124, 212, 226,
  227, 229
simultaneous functions 142
SIQR 242, 243, 249
skew 238
sort 39, 61, 96, 173, 179, 184–187, 197,
  235, 277, 285
spatial models 1, 9, 151
spreadsheet 19, 33, 106, 124, 129, 131,
  132, 134, 136, 140, 142, 144, 147, 172,
  177, 218, 219, 220, 237, 238, 243, 249
square grid 10, 152
  Cartesian 10, 168
stacks 170, 181, 184, 185
standard deviation 96, 220, 247–249, 265
statements 3, 7, 14, 51, 52, 60, 171, 259,
  275, 277–284, 286, 287, 291, 295, 296
statistical models 1, 17–19, 96, 101
statistics 18, 19, 133, 220, 230, 232–234,
  242, 244, 245, 247–249, 263, 265, 266
strict logic 287, 291, 298
string 7, 61, 82, 178, 179, 184, 185
subscripts 40, 82, 194
subsets 50, 52, 53, 68, 69, 75, 91, 261
substitution 225–227

subtraction 3, 4, 6, 87, 104, 105, 115, 116, 153
superscript 80
switching circuits 275, 284–286, 292
symbolic
statements 14, 277

taxonomy 16, 54
test 15, 16, 21, 24, 31, 32, 34, 35, 46, 49, 60, 63, 75, 106, 109, 187, 207, 209, 211, 212, 214, 258, 263, 265, 271, 276, 280
test data 15, 21, 24, 31, 46, 49, 63, 75, 212
testing 4, 17, 21, 50, 63, 95, 96, 104, 187, 193, 231, 280, 281
three-dimensional 9, 127, 151, 152, 177
time series 250, 251
trace pointers 195
transforms 150, 157, 159, 160, 165–167
translation 157, 158, 160, 167
tree 16, 41, 187–195, 197, 269, 270
model 6, 9, 31, 34, 43, 124, 150, 248, 257, 274
trend-analysis 18
trend-line 137
trends 18, 136, 233, 250

truncation 107, 206
two-dimensional space 10
two-state 15, 16, 81, 274–277, 287, 297
two-state logic 15, 16, 274, 276, 277, 287

uncertainty 19, 100, 145, 230, 233
diminishing 230
unique 8, 10, 15, 46, 68, 142, 180
uniqueness 15, 16
universal set 16, 49, 52, 54
universe of discourse 16, 48, 52, 53, 55, 140, 260, 277, 288
unknown 6, 9, 51, 127, 217, 222, 225–229, 260

variables 8–10, 12, 29, 30, 35, 61, 75, 105, 137, 174, 184, 217, 219, 220, 250, 252–254, 283, 287, 290, 292
variance 247, 248
vectors 37, 150, 153–155, 159, 160, 171, 181
combining 80, 154
Venn diagram 16, 48, 49, 54, 57, 60, 260, 277, 279, 288, 289, 292

XOR 60, 279, 284